PRINT AND SPECIFICATIONS READING FOR CONSTRUCTION

PRINT AND SPECIFICATIONS READING FOR CONSTRUCTION

Updated Edition

RON RUSSELL

WILEY

Published by John Wiley & Sons, Inc., Hoboken, New Jersey.
Published simultaneously in Canada.

For general information on our other products and services or for technical support, please contact our
Customer Care Department within the United States at (800) 762-2974, outside the United States at (317)
572-3993 or fax (317) 572-4002.

Wiley also publishes its books in a variety of electronic formats. Some content that appears in print may
not be available in electronic formats. For more information about Wiley products, visit our web site at
www.wiley.com.

Library of Congress Cataloging-in-Publication Data:

Names: Russell, Ron, 1953- author.
Title: Print and specifications reading for construction / Ron Russell.
Description: Updated edition. | Hoboken, New Jersey : Wiley, [2024] |
  Includes index.
Identifiers: LCCN 2023040655 (print) | LCCN 2023040656 (ebook) | ISBN
  9781394202553 (cloth) | ISBN 9781394202560 (adobe pdf) | ISBN
  9781394202577 (epub)
Subjects: LCSH: Building—Details—Drawings. | Buildings—Specifications. |
  Blueprints.
Classification: LCC TH2031 .R78 2024 (print) | LCC TH2031 (ebook) | DDC
  692/.1—dc23/eng/20231227
LC record available at https://lccn.loc.gov/2023040655
LC ebook record available at https://lccn.loc.gov/2023040656

Cover Design: Wiley
Cover Image: © Media Trading Ltd/Getty Images

Set in 11/13 pts and Interstate by Straive, Chennai, India

SKY10066295_013124

To my family for unquestioning support;
To Bob Graham Sr., Facilities Solutions Group, for leadership and the definition of quality;
To Roger Wilson, for practical applications and friendship;
And to my friends at PBK Architects for their support in drawings and illustrations.

CONTENTS

ABOUT THE COMPANION WEBSITE

This book is accompanied by a companion website:

www.wiley.com/go/printspecreadingupdatededition

The website includes:

- Drawing PDFs
- Specifications
- Tests and Keys

INTRODUCTION

This book is written to teach students with minimal exposure to the construction process how to find the design information in the drawings and specifications needed to work on a construction project. It is focused on the student learning about the development and use of these documents. This book is not about teaching the student construction or drafting, which is beyond our scope.

The book has two distinct sections. Section I is about teaching the basic information needed to understand the documents and how to read them. Section II is practical application of the information learned in Section I.

Section I is focused on giving the student a history of the drawings and specifications for a commercial construction project, and how they are developed for each project. This information sets the tone for the students' understanding of how they evolved and how they are used in the commercial construction process today. Section I then explains the structure of the documents and how to read them to find information. Quizzes and tests can be developed from the listed learning outcomes at the end of each unit.

Section II is focused on each category of the drawings and explicitly details what information is contained in each, and which divisions in the specifications pertain to that category. It is intended that Section II be used with a set of drawings and specifications along with a questionnaire, or list of questions for each category. This list of questions would be developed from a set of drawings and specifications for a commercial construction project from the local community and would require the student to acquire the answers from the documents much as they would in a construction environment for developing bids or solving field problems. A questionnaire would be developed for each category of drawings, as many questions as required to ensure student mastery of the information contained in a particular category and mastery of the process of using the documents structures to find information. Instead of providing the drawings, specifications, and questionnaires, it is recommended that the drawings and specifications for a local commercial construction project be used so that the student could actually visit the site and see the completed construction project as a learning aid. The questionnaires would be developed as requests for information from these documents, requiring the students to utilize their knowledge to find the answers as the desired learning outcome for each unit at www.wiley .com/go/printspecreadingupdatededition.

The book is structured to give the student three-point exposure to the information in each unit. The first point of exposure is having the student read the book. The second point of exposure is the students' listening to the lecture on the information they have read. The third point of exposure is practical application with the quizzes, tests, and questionnaires. This three-point exposure process of reading the text, hearing the lecture, and practical application ensures that the student has maximum opportunity to absorb the information and to be successful in reading the drawings and specifications on any commercial construction project.

SECTION I

THE PROJECT PROCESS

Section I will define the evolution of a commercial construction project with the team members involved and the contract documents developed during each phase of the project.

We will also learn about how the two primary contract documents, the working drawings and specifications, are developed and used.

Finally, we will learn a process for quickly finding information in these documents.

1

EVOLUTION OF THE CONSTRUCTION PROJECT

OVERVIEW

To better understand the working drawings and specifications, you need to have some understanding of the process that most construction projects go through from conception to completion. In this unit, we will look at the project team members and their responsibilities, and explore the sequential steps of a project through four phases: conception, promotion, design, and construction. As we explore these phases of construction, we will be referencing Figure 1.1. This graphical representation of the phases will allow us to tie together the project tasks, team members, and contract documents developed and used on most construction projects.

THE CONSTRUCTION PHASES

The construction of a building project goes through four distinct phases. Each phase results in the involvement of many different organizations contributing information to the building projects design. These phases and organizations ensure all of the necessary elements are addressed before completion of the project, and it is important that we understand them.

Print and Specifications Reading for Construction, Updated Edition. Ron Russell.
© 2024 John Wiley & Sons, Inc. Published 2024 by John Wiley & Sons, Inc.
Companion website: www.wiley.com/go/printspecreadingupdatededition

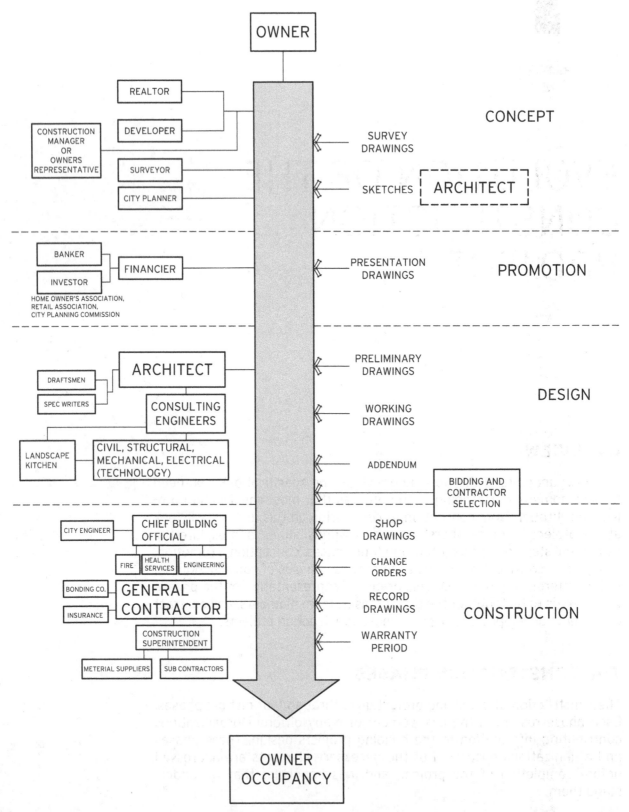

FIGURE 1.1 Four Phases of Construction

The Conception Phase

The start of the project comes when the owner decides a new building is needed. This decision is made in response to many possible factors. Perhaps the owner decides more manufacturing space is needed, more storage, or more warehouse space. The owner might decide he needs additional seating capacity to entice more customers to visit an establishment. Perhaps he just wants to have an updated facility or building that conveys the new image of the company. Whatever the reasons that motivate the owner, they result in the need for construction. The owner is the driver for the entire process since he has the design and the money to accomplish the project.

Depending on the owner and the business, he may or might not be very familiar with the construction process. If the owner constructs a lot of buildings each year, such as in retail or chain restaurants, he may be very familiar with the process and how to manage it. If he only builds every 10 to 20 years and building is not part of his business plan, he may be quite unfamiliar with the construction process. If the owner is not familiar with the process, it is probable at this point in the process that he or she would enlist the services of a construction manager. This individual would be well versed in the construction process and would advise and assist the owner through all phases of process. He would, in some cases, act as the agent for the owner and make commitments that bind the owner. In these cases, the construction manager could be called the owner's representative. If the owner builds a significant number of buildings each year, it is probable that there is someone on staff who will manage the process for the owner. In fact, if he builds many buildings each year, the owner might have a complete real estate and construction staff.

Once the owner has decided to build, he must then decide where to build. It may be that he already has property, especially if he has a campus-style setting for his company's operations. Or, he may simply possess property that he intends to use. If not, he would begin the search for property that would fit the business need or that would enhance the building's function. Most likely, the owner would begin the search by contacting a developer or a Realtor. This individual would direct him to the available properties that would most likely satisfy his business requirements.

A developer would be likely to show the owner properties that are undeveloped, lacking roads, utilities, and other amenities, but that would fit the owner's requirements and allow the flexibility to fully customize the site and building. The developer would generally offer up the property and agree to install the necessary roads and utilities to fit the owner's project requirements. These improvements would typically be developed as the architect and engineers develop the construction project. The advantages to this approach are that the owner has a site that fits his or her requirements exactly when completed, including all traffic requirements and with all utility requirements sized correctly for the finished project or building.

A Realtor, by contrast, would introduce the owner to properties that are already developed with roads or driveways and parking, all utilities available at least at the property line, and perhaps even an existing building that could satisfy the owner's requirements. The other significant difference between a developer and a real estate agent is that the developer is working with properties that he or she typically owns, whereas a real estate agent is showing properties that are owned by someone else.

Once the owner has selected a property for his project, he would want to make sure that the boundaries are defined and filed with the local municipality or county to ensure there is no confusion on where work can be done on the property. Most municipalities have set-back rules that govern how close a building or structure can be to adjacent property lines. The owner would also want to know where all easements are located for the municipality or utility companies. To accomplish the recording of the boundaries, a surveyor will be hired to come in and develop drawings and a written description of the property (Figure 1.2).

A surveyor will measure the distances around each boundary of the property, using sighting equipment, and document the information, using a metes and bounds system. He or she will note exact locations of corners, directions of turning points, and changes in elevation as necessary. The surveyor will document these dimensions by developing the survey drawings and a written description. These documents will be filed with the city or county the property resides in and will be recorded on the plat drawings for that municipality.

Simultaneously, the owner should be starting conversations with the appropriate city planners to determine if the property can be zoned for the type of activity the owner is going to use the property for, if it is not already so zoned. This is especially important for undeveloped properties because zoning may not have been established for the area yet and the process for obtaining zoning could be time consuming. Most properties shown by a Realtor would already be zoned and, if not zoned for the owner's specific activities, appropriate zoning must be obtained. The city planners can give the owner some idea of the amount of time and work that could be involved in this process.

Recognizing that the city planners will need some idea of the project to establish zoning requirements, the owner must now select an architect to develop the drawings and specifications. The architect will begin to be more heavily involved with the owner to start defining the project for which he or she has the primary responsibility for the design and development of the contract documents.

At this point, only the owner has any vision of what the project will look like. The architect must extract this vision from the owner and begin committing the vision to paper so others can observe the project as well. To do this, the architect will begin a series of meetings with the owner. The first meeting will be to gather basic data on the owner's needs, what type of business the owner will be conducting, how many people will occupy the building, how much raw material is needed at any given time, and so on. The architect will want to understand the owner's intended workflow and processes, or any people flow requirements that are required to satisfy the business intent. Next, the architect will begin to ask the owner questions about the vision

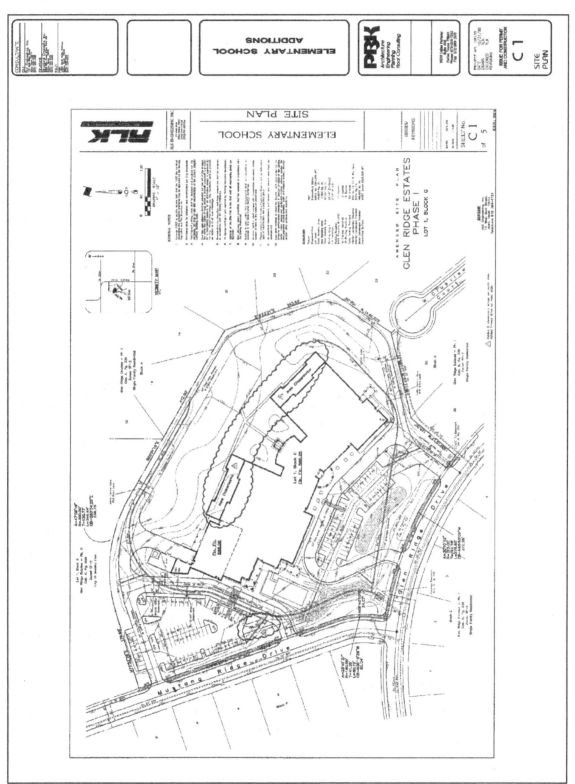

FIGURE 1.2 Survey Drawing

of the project: How does it fit on the property? Which direction does the building entrance face? How many room should the first floor have? and so on. With the data gathered at this first meeting, the architect can begin developing sketches based on an understanding of what the owner wants. This called the SD phase, or sketch development phase, by the architect.

At this point, the architect will focus on perhaps a basic floor plan, general site layout, and front elevation view of the proposed building. These sketches should reflect the comments of the owner on product or people flow, building appearance, and project's intended use. Once the architect has these first sketches developed, he or she will return to the owner and review them to validate the architect's understanding of the owner's vision.

The owner now has something to look at and can confirm for the architect that the sketches reflect his vision. Also, the owner can now see the project and apply changes. The architect might say, "Based on our last conversation, I have developed these sketches of how I envision your project." The owner would reply, "Yes this is what I want here and here; however, now that I see your sketch, I think this should really be over here." This type of exchange could continue for weeks or longer, if needed, until the architect is convinced that the same vision is shared by him and the owner.

During these exchanges, the architect would be providing guidance on things that are or are not feasible for construction. The architect would also provide insight on costs for certain elements and would guide the owner on design decisions that might be affected by codes and laws. The owner would finally reach a confirmation point: "Yes, that is exactly what I want." Then, the architect would begin developing what are called *presentation drawings* and the project would move into the promotional phase.

Promotional Phase

At this phase, many interested parties will need to understand the impact of the project, for many various reasons. The object of presentation is to get buy-in from the interested parties using the various forms of drawings.

Presentation drawings consist of several different types of drawings that will allow others to clearly see the project as defined by the owner and make decisions about how they would act on the project. The presentation drawings are designed to show the project as it will probably appear when completed. There are typically three types of these drawings: floor plans, pictorial representations, and models. You have probably seen examples of these yourself.

Floor plans are used a lot when an existing building is being expanded or remodeled. This type of drawing would be posted at the entrance of the building to allow the occupants and visitors to see how interior configuration changes will affect them and the work they do. Pictorial representations, as shown in Figure 1.3, and models, shown in Figure 1.4, are used primarily for new construction projects to allow interested parties to visualize the completed project. Pictorial representations are typically drawn using a perspective drafting technique and are usually colored or painted to add realism.

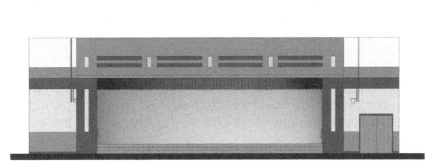

FIGURE 1.3 Presentation Drawings

FIGURE 1.4 Architectural Model

Models are a type of pictorial drawing that offer a three-dimensional view of the project made to scale. You may have been in lobbies of some buildings and seen a picture of the building on the wall or a glass-enclosed three-dimensional model on display. These drawings and models are often displayed after completion of the project when *interested parties* no longer need to be swayed for support. Interested parties could include financiers such as bankers or investors, city planners or engineers, homeowners associations, retail shopping centers, or mall associations.

When an owner decides to build, he or she might chose to finance the project, even if the funds are available for the construction. There are many different types of financiers, but two are most typical. *Bankers* are lenders that intend to make money on the interest charged the owner for the use of the money. They would be interested in seeing the presentation drawings to determine if the owner's project will suit his intended use and would be a viable asset over the life of the loan. *Investors,* by contrast, will lend money on the project in anticipation of getting a return on their investment from the valuation of the project over an extended period of time, either from the operations conducted at the site or from the building's use through leases.

As mentioned previously, city planners are interested in the project for zoning, business area planning, and code compliance standpoints. They are chartered by the municipal government to ensure that all applicable building codes are complied with by all new and renovation construction projects and the places are safe for the general public to conduct business. They are also concerned with assisting the municipality with planning business areas so that a certain compatibility exists between surrounding businesses and residential areas. This often involves modifying zoning requirements for particular pieces of property. If you are looking at zoning around a residential area, you should be prepared to show your presentation drawings to homeowners associations.

Homeowners associations are interested in preserving the quality of life that their areas represent. They are typically chartered by the neighborhood in their area to monitor anything that has the potential to change that quality of life and bring it to the attention of all the members of the association. As a group, they would be interested in how a new project will affect home and property values, if it will fit in with the neighborhood's appearance, if it will bring services that the community needs, if it will have an adverse affect on traffic, and so on. They will tend to scrutinize every aspect of the project, including exterior colors, to ensure that their property values and quality of life are not degraded.

Much of the same is true of retail shopping center and mall associations, only from a business perspective. These interested parties have committed their livelihood to a certain location and will be very concerned that any new project might jeopardize their business by not meeting the standards established by the retail center.

Many interested parties will be looking at the proposals made for the project. The owner and architect will be required to work together closely to satisfactorily allay concerns over the project. Clear and descriptive presentation drawings go a long way toward accomplishing this, because many people have difficulty in visualizing a project without a representation of it. One picture could indeed be worth a thousand words in these situations.

Design Phase

Once the promotional phase is significantly passed, the architect and his engineers begin in earnest on the DD phase, or design development phase. In this phase, the architect has provided the primary consulting engineers with the basic premise of the design, and they begin to work on the project's preliminary drawings.

The preliminary drawings are drawings created by the architect and engineers to solve problems. These drawings will be used to make sure that the design intent for the building can be met. For example, they will be used to solve problems in areas where dissimilar materials come together, or where unique building shapes require unique structural members. All problem areas will be explored and some type of solution developed before progressing on the project.

The architect's primary consulting engineers for electrical, mechanical, and structural issues will work with the architect to solve each respective discipline's problems. The electrical consulting engineer will work to solve any special lighting issues, power distribution issues, or dedicated electrical service issues and will provide the solutions to the architect in drawing form, as well as with the appropriate specifications for the electrical engineer's portion of the design. The mechanical consulting engineer will develop solutions for heating, ventilating, and air conditioning (cooling) (HVAC) distribution to balance the building environment and ensure that systems such as supply water and fire protection can be completed as required by code. The structural engineer's specification writers will make sure that the building design can be supported structurally and will withstand the loads anticipated during the conducting of business in the building and from the outside environment. Once the preliminary drawings are completed, the consulting engineers will provide them to the architect for review. If at any time during the development of these drawings a problem cannot be solved, then the architect must alter the design so that the project can be accomplished. This may necessitate going farther back to the earlier phases of the project with the owner and interested parties if the design must be significantly changed. Once all issues have been resolved, the architect and consulting engineers can begin developing the drawings that we are most interested in, the working drawings. The architect will call this the CD phase, or contract document phase, which the working drawings will become.

The working drawings are the drawings that we typically bid from and build from, hence the term *working* drawings. Once these documents are completed, they are made available to contractors for bidding—either through a plan room where they are checked out and reviewed or by direct issue, where the contractors all pay a deposit to the architect for their own copy. At the same time, the specifications are issued so the interested parties have a complete picture of the project to bid on. The specifications have the invitation to the bidders, instructions for the bidders, and the bid forms to be used for that project. During the bidding process, questions may arise from the bidders and other interested parties that require information in the working drawings and specifications to be changed. These changes will be issued in the form of an addendum, which modifies the working drawings and specifications by adding information to or by deleting information from these documents. Addenda will be issued to all interested parties who have solicited drawings from the architect and to all plan rooms where drawings and specifications have been submitted. The addenda will only be issued up to about two to three days before the bids are due. This date will be announced in the bidding instructions. Once the bids are submitted, no more addenda will be issued or modifications to the working drawings and specifications made.

Construction Phase

Once bids are received and opened, usually the lowest qualified bidder is selected. The qualifications required for bidders on each project can vary. Once selected, the lowest qualified bidder will become the general contractor for the project. After contract documents are signed, the general contractor will go to the city offices and obtain a building permit. This will set up the city building inspectors to develop an inspection schedule for the contractor for the various stages of the work, which will have to be approved before work on each subsequent stage of the project can continue. Typically, the city inspectors will include the plumbing inspector, electrical inspector, structural and fire protection inspectors, and possibly the city engineers, as needed. These inspections should go smoothly since the city will have been involved early in the project and will have seen the various drawings created at each stage: presentation, preliminary, and working.

The general contractor is responsible for providing all labor, materials, and equipment to complete the project as defined by the working drawings and specifications. To provide the materials, the general contractor will work with suppliers' and manufacturers' representatives, using the specifications to obtain exactly the materials specified by the owner and the architect.

Because most general contractors do not keep all the crafts and skilled labor on staff to complete and entire project, they will use subcontractors with their own workers to provide the labor for the project. Most subcontractors tend to be specialized in one particular aspect of construction and have the necessary specialized equipment for completing that part of a project. Subcontractors are sometimes also responsible for providing drawings of the work they are going to do. These are called *shop drawings*.

Shop drawings are created by a contractor or subcontractor that show details of the work that they are going to complete for the project. These drawings are typical of elements of the building that will be fabricated in the contractor's or subcontractor's shop—thus the name *shop* drawings—and brought to the job site and put in place. These drawings must be submitted to the general contractor who, in turn, submits them to the owner, or architect if the designated recipient, who reviews them for compliance to the working drawings and specifications, approves them if acceptable, and returns them to the general contractor, who gives them back to the subcontractor to begin work. Work on these elements of the building should not be started until the shop drawings are approved.

All work done by the subcontractor must be completed as indicated on the shop drawings.

During the construction phase, things might appear that trigger a change in the scope of work defined in the working drawings and specifications. Perhaps something is uncovered that was not seen before or anticipated. Sometimes during the course of construction, a design simplification becomes evident. Sometimes, an error is found that will not allow the project to proceed as intended. Any of these events will require that the architect and consulting engineers devise a solution to the new problem. If this solution necessitates changing the scope of work and modifies the working drawings and specifications, a change order may be issued.

Change orders modify the contract for construction, and possibly the contract amount, by adding information to the working drawings and specifications, or by deleting information from the working drawings and specifications, thereby modifying the scope of work. This is very much like the effect of an addendum; however, it is much more significant because we are now modifying a scope of work that has a contract associated with it for a specified amount of money for that scope of work.

As these changes arise during the construction process, they must be recorded. This is typically done on a copy of the working drawings kept in the job superintendent's office at the job site. When a change is encountered due to a change order or for some other reason, the modifications showing how the construction was actually done are marked in red pencil or ink on the drawings and in the specifications. These marked-up drawings become the basis for the *record drawings*. Once the project is completed, the marked-up drawings are submitted to the architect.

The record drawings, or *as-builts*, as they are commonly called, document for the owner and contractor exactly how the project was built and where elements of the building were located. The architect modifies the original working drawings and specifications to provide a record of how the project was actually built so the owner will be able to use the documents for maintenance, repairs, and modifications to the building in the future.

The contractor is interested in these record drawings for another reason. Most general contractors will provide the owner with at least a year's warranty on the labor and materials provided, including workmanship. In some cases, contractors have began to use the warranty as a competitive advantage by offering extended warranties, up to five years. It is then important that the contractor have a record of exactly how the project was completed in order to effectively manage that process. At the end of the warranty term, the project can then be considered completed.

We have now looked at the four phases of the constructions process, noted some of the individuals involved with the project, and learned where each of the different categories of drawings fit into the process. Next we will look at the relationship of some of the main construction team players and explore in more detail their responsibilities.

THE CONSTRUCTION PROJECT TEAM MEMBERS

There are three primary members on the traditional construction team: the owner, architect, and general contractor. Each of these team members has various individuals or groups of individuals that support them in the delivery of their part of the project. They all have specific roles and responsibilities to each other. Although there are other individuals that might be on a project team, we need to understand the primary members first.

The Owner

Since the construction process begins with the owner's identification of its need for a facility, we will start with learning the owner's role in the construction process. Without the need for a building, we would not do construction and the owner would be saving money. But since the owner is spending his money, he plays some critical roles. Also, the role of spending money does not alleviate the owner's responsibilities to the other team members. It is probably appropriate to point out that even though we are saying *owner*, it is possible that the actual team member carrying out the duties could actually be a construction manager or an owner's representative.

The owner obviously has responsibility for paying the other team members for services rendered. The architect is paid upon completion of various design stages, and the general contractor through pay applications that are submitted monthly, or on some other agreed-on frequency. The pay applications indicate a certain amount of completed construction work or materials for the project that are staged on the site. Once the owner has verified the accuracy of the contractor's pay application, sometimes with the help of the architect, the owner has a specific agreed-upon time frame in which to furnish payment to the contractor.

During the design and construction phases of the project, the owner will have the responsibility for providing information to the architect and general contractor within specified periods of time to ensure that the project remains on schedule and that no one incurs unnecessary expenses due to delays. These duties will be outlined in the contractual documents signed by these parties and discussed in the next chapter.

The owner also must provide the construction site without encumbrances so the work can be accomplished. The general contractor must be able to utilize the site effectively to allow safe and cost-effective construction, within certain physical limitations such as right-of-ways, easements, and property boundaries. The owner is also responsible for providing the description of the project through the drawings and specifications, even though these are created by another team member, the architect.

The Architect

The architect's primary function is to develop the definition of the project and provide the final construction documents, the working drawings and specifications. We have talked at length about the conceptual, promotional, and design phases of the construction process, and this is where the primary responsibilities of the architect are. The architect will assist the owner in developing the concept, refining it, presenting it to the interested parties, and developing the final construction documents. The architect is usually the primary individual who works through the first phases. Once the project moves into the design phase, the architect will enlist the aid of other members of the team to complete the working drawings and specifications.

To complete these contract documents, the architect will typically enlist the services of structural, mechanical, and electrical consulting engineers. Usually, unless the architect has a very large firm, these disciplines will not be on staff. Instead, because requirements vary from project to project, these individuals are kept on retainer to assist as needed on each project. These engineers will complete their portion of the project design based on information received from the architect for assimilation into the working drawings. These engineers will also be responsible for developing the portions of the specifications applicable to their design. The architect has the overall responsibility for the integrity of the finished working drawings and specifications.

We have noted that the owner or the owner's representative will exercise the owner's duties. On some projects, the architect will assume the owner's representative's duties at the owner's direction and as stipulated in their agreements. This arrangement is why it sometimes appears that the architect is the main operative in the team, when, in fact, the owner is always the primary member coordinating the architect and general contractor's efforts.

The General Contractor

The general contractor is responsible for completing the project as defined in the working drawings and specifications and for developing the means and methods for accomplishing the work, scheduling the work, and providing all labor, materials, equipment, and supervision necessary to complete the project.

Much like the architect, the general contractor has many varying projects and requirements. As a result, the general contractor will not always keep skilled labor or specialty equipment on hand. Instead, he will utilize subcontractors who have the skilled labor and equipment needed to complete certain portions of the project. The advantages of this for the general contractor are as follows:

- It reduces overhead by not having a significantly large number of pieces of equipment and skilled labor on hand that would not be fully utilized.

- The use of subcontractors ensures that skilled craftsmen are on the project.

- Since most subcontractors specialize in their fields of construction, they have the proper equipment to do their work.

For example, if all you do is electrical work, then it makes sense to have skilled and licensed electricians and the electrical construction equipment needed for most jobs. However, the general contractor could not cost-effectively provide this unless it were a large firm that continually utilized these in daily business on projects.

Since the general contractor is responsible for furnishing all materials, the general contractor will have rapport with various suppliers of construction materials. Depending on how often and how much of these materials are purchased from a particular vendor, the general contractor could have a competitive advantage over another contractor in supplying materials, as long as the materials provided comply with what is defined in the specifications.

The general contractor also has the responsibility for providing information in a timely manner to the owner regarding schedule progress, completed work, and any other issues that affect the project.

The team relationship between the owner, architect and general contractor is just that–a *team* relationship. The owner needs the project for his or her business. The architect wants to complete the project for payment of services. The general contractor wants to complete the project for payment of services. Each member of the team has a vested interest in the same objective–completing the project. Even so, the owner is the customer, and the team relationship reflects this.

THE CONSTRUCTION TEAM RELATIONSHIPS

To look at the construction team relationships, we will really be discussing the different types of construction contract delivery methods. There are primarily three methods to explore: the traditional, construction management, and design-build delivery methods.

The Traditional Method

The traditional method is the oldest and most commonly used of construction delivery methods. Shown in Figure 1.5, this type of team structure has the basic owner/architect/general contractor relationships and the services provided are as we have discussed previously.

The owner provides the money and the site for the project and has an agreement with the architect for providing the design, working drawings, and specifications. The general contractor has an agreement with the owner to provide the labor, materials, equipment, and expertise to complete the project per the working drawings and specifications. These agreements are specific written contracts that exist between the parties and will be discussed more in the next chapter. It is important to note that there is not a contractual agreement between the architect and general contractor. These individuals work exclusively at the discretion of the owner, based on their contracts with him, to complete the project.

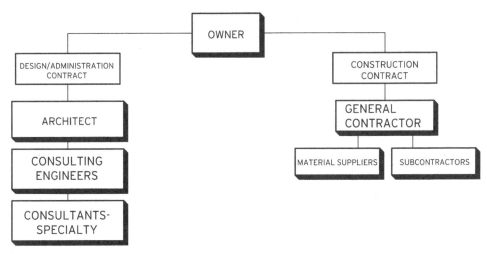

FIGURE 1.5 Traditional System

If you will notice in the example, those individuals who support the architect and general contractor are not considered primary team members. There are no contractual agreements between these individuals and the owner. The structural, mechanical, and electrical consulting engineers work exclusively for the architect and have no relationship to the owner or general contractor. The subcontractors and suppliers work exclusively for the general contractor and have no relationship to the owner or architect. This traditional team structure also typifies the information flow. Again, each team member has the same goal—complete the project.

The Construction Management Method

As discussed previously, there are many construction projects where the owner has no construction expertise, does not have the time to manage a construction project, or does so much construction that it makes sense to not simply have an owner's representative or architect perform those duties, but requires an entirely separate firm to be hired just to manage the project(s). This is also sometimes done when a project is very complex. The construction management firm will assume the roles of the owner and act as his or her agent to make decisions in directing the architect and general contractor to complete the construction project, as depicted in Figure 1.6.

The owner will hire the construction management firm, and it will make all the agreements with the architect for development of the construction documents to satisfy the owner's needs and then will manage the bidding and construction process to complete the project. Essentially, the other team members will operate with the construction management firm to complete the project.

There are two primary construction management delivery systems, agency and at-risk. The agency construction manager works for a fixed

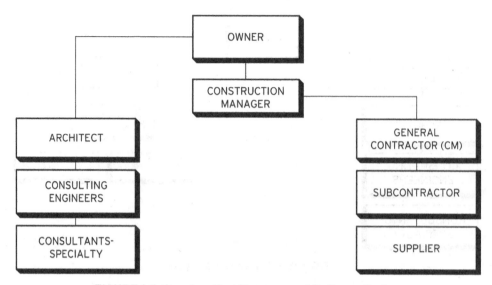

FIGURE 1.6 Construction Management Delivery System

fee with the owner and acts only as the owner's agent and manages the archi-tect and general contractor. The agreements between the owner and architect and general contractor remain the same as in the traditional method. The agent has no responsibilities for the design integrity or construction methods or techniques. The owner pays the agency construction manager on a fee basis.

The construction-manager-at-risk has similar responsibilities, except for being involved in the conception and design phases of the project, and manages the project as though he or she were the general contractor. The owner has little day-to-day involvement, and the general contractor is in a role similar to that of a subcontractor. The construction-manager-at-risk provides the owner with a maximum price for the project and manages the project so that the costs allow for completion of the project, with some profit left over for the manager-at-risk's organization— hence, the *at-risk* title. If the project is not managed well, costs will rise and possibly exceed the maximum price, and the construction manager will lose money.

Design-Build Method

This method is becoming more popular with owners, since it is a single-source project delivery method where all the necessary services are provided by one organization. Design and construction, even site selection and purchasing options, are provided to the owner. A single design-build organization will be responsible for providing the functions required to develop the design and the final working drawings and specifications, and for actually providing the construction of the project, usually with its own labor, although some of the project could be subcontracted to others. For the owner, there is a guaranteed maximum price. Figure 1.7 shows the relationship of the design-build firm to the owner.

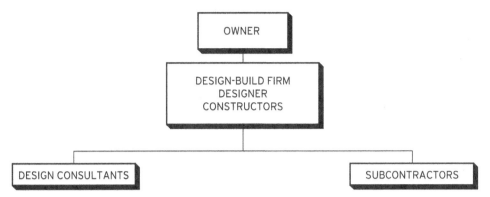

FIGURE 1.7 Design-Build System

SUMMARY

In this chapter, we discussed the four different phases of a commercial construction project: conception, promotion, design, and construction. We also discussed the individuals involved in the phases and their roles. We reviewed the responsibilities of the construction team members, owner, architect, and general contractor, and examined the project delivery methods that can be used to complete the project.

All of the relationships, project tasks, and construction methods mentioned here require a significant amount of organization and documentation. As a result, there are many documents that form the agreements for construction. No single document is inclusive of everything necessary to complete a construction project. In Chapter 2, we will explore some of the contract documents necessary to describe the agreements and manage the work.

▓ CHAPTER 1 LEARNING OUTCOMES

- The four project phases are: conception, promotion, design, and construction.

- The seven types of drawings created during the project phases are: survey drawings, sketches, presentation drawings, preliminary drawings, working drawings, shop drawings, and record drawings or as-builts.

- The three construction contract delivery systems are: traditional system, construction system (agency and at-risk), and design-build system.

▓ CHAPTER 1 QUESTIONS

1. How many phases are in the construction process? Describe what the role is for each phase.

2. List the seven types of drawings created during the construction process and what they are used for.

3. Describe the three basic delivery methods used to contract for and construct buildings.

2

CONSTRUCTION PROJECT CONTRACT DOCUMENTS

OVERVIEW

As mentioned in the previous chapter, the construction project is usually of such a magnitude and complexity that no single contract document can include all of the agreements or information necessary to manage the work and complete the project. There are several team members with multiple responsibilities, many definitions of the scope of work, and all to be effected over a lengthy period of time, thus requiring that many contract documents be used to form the complete agreement for construction of a commercial building project.

Even though there are many contract documents that could be used on a given project, in this chapter we will look only at the primary ones used on most construction projects. We will focus on the agreements between the primary team members, the documents that govern the construction process, and the drawings.

AGREEMENTS BETWEEN THE PRIMARY TEAM MEMBERS

There are two primary agreements between the team players that define each player's responsibilities in accomplishing the requirements of the owner: the *Standard Form of Agreement between Owner and Architect* and

Print and Specifications Reading for Construction, Updated Edition. Ron Russell.
© 2024 John Wiley & Sons, Inc. Published 2024 by John Wiley & Sons, Inc.
Companion website: www.wiley.com/go/printspecreadingupdatededition

the *Standard Form of Agreement between Owner and Contractor*. These agreements guide each team member in the execution of their duties and are the formal agreements between the members.

Agreement Between Owner and Architect

The architect's duties basically involve two primary tasks:

1. Development of the project design and construction documents, the drawings, specifications, and addenda

2. Administration of the project

Standard Form of Agreement Between Owner and Architect, AIA (American Institute of Architects) Document B101, outlines the standard architect's services for accomplishing those tasks (see Figure 2.1).

The first table of articles includes descriptions and definitions of responsibilities of the owner and architect and the scope of the architect's services and how the architect will be compensated. The second table of articles addresses the architect's design, project administration, construction procurement, and contract administration services.

This agreement will be used to describe the design services the architect will provide, and will also be used to define what, if any, responsibilities he will have in administering the project for the owner. As we have stated before, if the owner is not going to manage the project itself; he or she could contract with the architect through this agreement or utilize a construction management firm.

Agreement Between Owner and Contractor

The general contractor's responsibility is to provide the materials, labor, equipment, and construction expertise necessary to complete the project as described in the contract documents. The AIA Document A101, *Standard Form of Agreement Between Owner and Contractor,* shown in Figure 2.2, sets forth the articles for the scope of work and how the contractor will be paid.

Article 1 outlines all the contract documents that will be considered as a part of the agreement for construction. Article 2 defines the scope of work. Article 3 defines the date of commencement for the project and how substantial completion of the project will be determined. Articles 4, 5, and 6 define the contract sum, how progress payments will be processed, and how the final payment will be made.

This agreement will be the formal agreement that the owner and general contractor sign for the contractor's services. However, as with the agreement with the architect, there are other documents that describe the work and that all the parties will follow to complete the project.

PROJECT DEFINITION DOCUMENTS

Several contract documents used by all the project team members define the project and the processes that will be used for managing the project during construction. Some of these documents are in written form

AIA® Document B101™ – 2007

Standard Form of Agreement Between Owner and Architect

AGREEMENT made as of the   day of
in the year of
(In words, indicate day, month and year)

BETWEEN the Architect's client identified as the Owner:
(Name, address and other information)

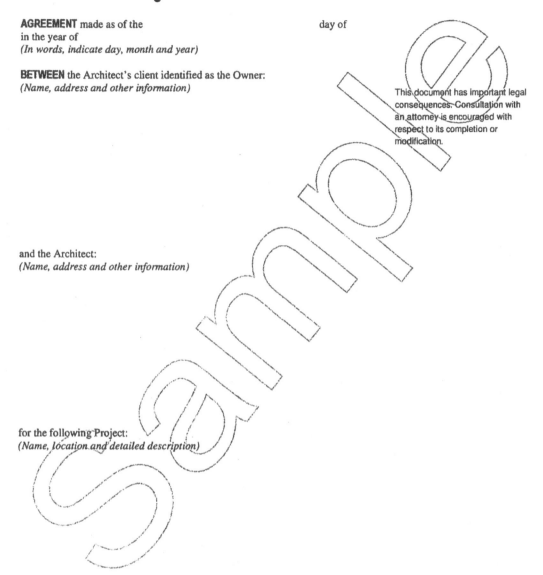

This document has important legal consequences. Consultation with an attorney is encouraged with respect to its completion or modification.

and the Architect:
(Name, address and other information)

for the following Project:
(Name, location and detailed description)

The Owner and Architect agree as follows.

FIGURE 2.1 *Standard Form of Agreement Between the Owner and Architect,*
AIA Document B101-2007

▓AIA® Document A101™ – 2007

Standard Form of Agreement Between Owner and Contractor where the basis of payment is a Stipulated Sum

AGREEMENT made as of the   day of
in the year
(*In words, indicate day, month and year*)

BETWEEN the Owner:
(*Name, address and other information*)

This document has important legal consequences. Consultation with an attorney is encouraged with respect to its completion or modification.

AIA Document A201™–2007, General Conditions of the Contract for Construction, is adopted in this document by reference. Do not use with other general conditions unless this document is modified.

and the Contractor:
(*Name, address and other information*)

for the following Project:
(*Name, location, and detailed description*)

The Architect:
(*Name, address and other information*)

The Owner and Contractor agree as follows.

FIGURE 2.2 *Standard Form of Agreement Between Owner and Contractor,*
AIA Document A101-2007

and some are in graphical form, or drawings. These documents consist of the working documents, general conditions, specifications, addenda and change orders.

The Working Drawings

The drawings used for the construction project are called the working drawings because they are worked on to bid and build from. They are developed by the architect and belong to the architect. They convey the graphical definition of the project and are used by the general contractor, subcontractors, and suppliers to construct the project.

General Conditions of the Contract for Construction

The *General Conditions of the Contract for Construction*, AIA Document A201-2007, shown in Figure 2.3, brings together the tenants for the team members set forth in the agreements between the owner, architect, and general contractor. It also defines the rights, privileges, and responsibilities of the team members, and provides overall direction for the administration of the project. This document originated in the early 1900s and has evolved to its current form over the years. As a result, it contains articles used on most commercial construction projects. It is typically used unchanged, or with very minor modification, since the articles contained within apply to most projects. Later, we will look at how they are modified for a specific project.

General Conditions of the Contract for Construction contains the following 15 articles:

1. General Provisions
2. Owner
3. Contractor
4. Architect
5. Subcontractors
6. Construction by Owner or by Separate Contractors
7. Changes in the Work
8. Time
9. Payments and Completion
10. Protection of Persons and Property
11. Insurance and Bonds
12. Uncovering and Correction of Work
13. Miscellaneous Provisions
14. Termination or Suspension of the Contract
15. Claims and Disputes

▲AIA® Document A201™ – 2007

General Conditions of the Contract for Construction

for the following PROJECT:
(Name and location or address)

THE OWNER:
(Name and address)

THE ARCHITECT:
(Name and address)

TABLE OF ARTICLES

1   GENERAL PROVISIONS

2   OWNER

3   CONTRACTOR

4   ARCHITECT

5   SUBCONTRACTORS

6   CONSTRUCTION BY OWNER OR BY SEPARATE CONTRACTORS

7   CHANGES IN THE WORK

8   TIME

9   PAYMENTS AND COMPLETION

10   PROTECTION OF PERSONS AND PROPERTY

11   INSURANCE AND BONDS

12   UNCOVERING AND CORRECTION OF WORK

13   MISCELLANEOUS PROVISIONS

14   TERMINATION OR SUSPENSION OF THE CONTRACT

15   CLAIMS AND DISPUTES

FIGURE 2.3 *General Conditions of the Contract for Construction,*
AIA Document A201-2007

These articles are basically the same for each project and are agreed to by each of the team members. The general conditions are typically included in the specifications by reference and are modified by a section in Division 1 called Supplemental Conditions.

The Specifications

The specifications are the primary written contract document for a commercial construction project and have evolved over the years. This document, very much like a book, is unique to each building project and is issued with the set of plans by the architect. The current format was developed by the Construction Specifications Institute (CSI; see www.csinet .org). This format breaks the written description of the project into project divisions. Originally, there were 16 divisions—with one division, Division 0, added that contained all of the bidding information. Then a Division 17 was added that contained all of the low-voltage systems and removed those systems from Division 16, Electrical. Eventually, the specifications were expanded to include 49 divisions, which broke the information into more distinct systems and was intended to make the information easier to find. This new format has been slow to catch on in the construction industry with companies that typically deal with midsized to smaller projects, but it is gaining ground. Of the 17 project divisions, Division 1 is the nontechnical division and Divisions 2 to 17 are the technical divisions that describe the specific written requirements for the construction of the project. The specifications include many other important pieces of construction information that pertain to the project.

BID DOCUMENTS

The bid documents are typically included in what is called Division 0. These documents include all the instructions for submitting bids for a particular project. This bidding information includes the invitation to bid, instructions to the bidders, and the bid forms. These documents will contain information only for the project in whose specification book they are included. This information will be different for each construction project.

The invitation to bid contains information about the project, the owner and architect, who is invited to bid, and when and where the bids must be submitted. The instructions to the bidders contain all the information needed to correctly complete the bid forms and properly submit them. These instructions will define the bonding requirements, contractor qualifications and disclosures, and how to complete the forms. The bid forms must be completed exactly as indicated, including certification that all addenda have been considered in the amount being submitted to complete the project. The forms shown in the specifications must be used to submit the bid. No other forms will be accepted, and these forms will only pertain to that specific project. The bidding requirements and submittal date and time must be met exactly as indicated. Any deviation will result in the bid being disqualified.

DIVISION 1 GENERAL REQUIREMENTS

This division contains what are considered the nontechnical specifications of the project and outlines the general requirements that will be used to govern the other divisions. This information will vary from project to project, but it will typically include things such as requirements for the site project signs, project clean-up, start-up of new equipment, and clean-up of the site. Also included will be the supplementary conditions.

As mentioned previously, the *General Conditions of the Contract for Construction* is purchased off the shelf and is typically used in its entirety for the construction project with little or no modification. However, it is usual, because of each construction project's unique requirements, to include supplementary conditions to make significant modifications or additions to the general conditions. These supplementary conditions will reference the AIA document and will create conditions and requirements that are specific only to that particular project and these modifications will be located in Division 1, Supplementary Conditions.

OTHER DIVISIONS

These divisions are considered the technical specifications and delineate the specific requirements for the materials that are allowed on the project, the quality controls for the construction process, and the methods allowed in protecting the work. Also included will be many documents that will be referenced, but not printed in the specifications, called *reference documents*.

It is not uncommon for specified requirements to be derived from another organization's written standards. The specifications will note that some aspect of the work must be executed as specified in the document created by the originating organization. However, instead of including that entire document or the appropriate section applicable to this project, it is included in the specifications by reference. If work is specified to be completed as defined in a reference document, the contractor must complete the work as though the definition were printed in the specifications for the project. A contractor who is not sure of what the referenced document says about that aspect of the work must obtain a copy of that document and complete the work as indicated.

Addenda and Change Orders

After the construction documents (working drawings and specifications) are issued for bidding, there will be requests for clarifications. Some information in these documents might need interpreted; the intent of other sections might need to be explained. Sometimes there will be a prebid meeting on the construction site, where all interested bidders can meet the owner and architect and ask questions about the project documents. During the course of this meeting and the bidding process, questions will be asked and the owner and architect will provide answers in the form of addenda. These addenda will either add information to, or delete information from, the working drawings and specifications. The architect will have kept track of all documents issued for bidding and will mail the addenda to each registered bidder. As already noted, bidders must declare on their bid forms that they have seen all the addenda. Any bidder that does not do this will be disqualified.

The number and amount of information contained in addenda will vary for each project. Addenda will only be issued up to a few days before the bids are due. Then no more information will be provided. Bids submitted will be in consideration of the working drawings, specifications, and all issued addenda. These documents will not be changed again until after the contract has been awarded to the lowest qualified bidder. These modifications will be in the form of change orders.

After a contract has been executed between the owner and contractor, a stipulated sum will be established for the scope of work defined in the working drawing, specifications, and addenda. Any modification of this contract agreement must be made through the change order process. Changes in the information in contract documents will still have to be modified to allow for unforeseen issues, omissions, and changes in the owner's requirements, thus modifying the contract amount. Change orders are issued to add or delete this information in the contract documents. The contract amount will also be modified as a result of the change in the scope of work. If a change is required, the owner will have the architect create documents stating the changes needed in the scope of work. These changes are given to the contractor to price out—changing the contract amount up or down, depending on the change in scope of work. This change in amount is submitted to the owner, who has the option of agreeing to the contract change or seeking another contractor to complete the work. If the owner agrees and signs the change order, it becomes a contract document for that project.

SUMMARY

In this chapter, we have looked at the various documents involved in constructing a commercial building. The agreements between the team members included the *Standard Form of Agreement Between Owner and Architect* and the *Standard Form of Agreement Between Owner and Contractor*. The project definition documents included the *General Conditions of the Contract for Construction*, the specifications, and the working drawings. Examples of these documents are shown at the end of this unit.

The contract for construction on a commercial construction project comprises all of the documents defined in this chapter. Each document plays a specific part in providing information for the completion of the project. Even though there are many documents involved, we are primarily interested in the specifications and working drawings. We will explore these documents in more detail in the next two chapters.

■ **CHAPTER 2 LEARNING OUTCOMES**

- There are separate contracts between the owner and the general contractor, and the owner and the architect.

- There is no contract between the architect and general contractor.

- The drawings and specifications are the primary contract documents for defining the project.

■ **CHAPTER 2 QUESTIONS**

1. How many contract documents make up the contract for construction?

2. At any time during the construction process does a contract exist between the architect and general contractor? Why do you believe this is the case?

3. What are the primary contract documents defining a project?

3

WORKING DRAWINGS

OVERVIEW

In this unit, we will explore the process for creating the working drawings and their structure so that we can understand how to move from sheet to sheet and find the information for bidding and building the project, which are the drawing's' primary uses. We will occasionally use the term *drawings* to mean working drawings.

As stated previously, these documents are created by the architect for the owner and are used by the general contractor, subcontractors, and suppliers to complete the project. Now we need to look at the techniques the architect, engineers, and drafters use to construct the documents so we can better understand how information is placed on them. Our intent is to understand this structure so well that minimal time is spent needlessly thumbing through the drawings to find information.

One of the first pieces of important information to learn to minimize time spent in searching for information is that the drawings show us *size*, *shape*, and *location*. These are the documents that provide us the graphical representation of the project and give us the basic three-dimensional description of the project to construct.

THE DRAWING STRUCTURE

The drawing structure is unique to the architect who creates the drawings. However, there are certain standard conventions that almost all adhere to and those conventions exist even though they might be called by another name. So, if we learn to recognize what these conventions are, we can recognize them in whatever form or by whatever name they might be called.

Print and Specifications Reading for Construction, Updated Edition. Ron Russell.
© 2024 John Wiley & Sons, Inc. Published 2024 by John Wiley & Sons, Inc.
Companion website: www.wiley.com/go/printspecreadingupdatededition

Drawing Categories

The first thing we need to learn about the working drawings structure is that all the sheets of a project drawing package will be categorized, or grouped, into six primary categories.

1. Civil

2. Architectural

3. Structural

4. Mechanical

5. Electrical

6. Plumbing

All working drawing packages will typically have these six categories, and all the drawing sheets will be placed into one of these categories, depending on the type of information contained on each individual sheet.

There can be other categories in some sets of plans, depending on if certain information is of a critical nature, or if the amount of information is so great that it warrants a separate category. Examples of this might be landscaping or fire protection drawings. The landscaping drawings are typically included in the civil drawings, but if there is a significant amount of landscaping to be completed on a project, or if the description of the landscaping, as related to size, shape, and location, is significant enough, then those drawings', information might be separated into another category of its own titled "landscaping," and would be created by the landscape architect. The same is true of the fire protection information. This information could be included in the mechanical category. However, if the fire protection system for a project is complicated or significant, it can be made into a category or group of drawings of its own. On many projects, this is required because many municipalities specify that these systems be engineered and installed by a licensed fire protection company. We should also mention the potential combining of the plumbing drawing information. This information, on some commercial construction projects, could be included in the mechanical category. However, with most commercial construction drawing sets, it will be placed in a separate category of its own, called "plumbing." For our purposes, we will only be considering the six primary categories and the information contained in each of those categories.

To make sheet identification easier within each category, the drawing sheets in a category will be numbered to begin with, or include, the letter of the category it belongs to. For example, civil drawing sheets will be numbered so that they include the letter C. Architectural category drawing sheets will be numbered so that the sheet number includes an A. The structural category drawing sheets will have numbers that include an S, and, the electrical and mechanical categories drawing sheets will have numbering systems that contain an E and M, respectively. This allows for easy determination of where a drawing sheet fits into the working drawings if the drawing is found by itself. There are numerous drawing numbering systems that can be used, but most will conform to this practice, see Figure 3.1. We will look at the different types of numbering systems available later.

SHEET INDEX

C1	SITE PLAN
C2	SITE GRADING PLAN AND DETAILS
L-1	LANDSCAPE PLAN
L-2	IRRIGATION SYSTEM PLAN
A3	FLOOR PLAN
A4	EXTERIOR ELEVATIONS
A5	INTERIOR ELEVATIONS
A6	WALL SECTIONS
A7	SCHEDULES-WINDOW AND DOOR DETAILS
A8	REFLECTED CEILING PLAN AND DETAILS
S-1	FOUNDATION PLAN AND DETAILS
S-2	ROOF FRAMING PLAN AND DETAILS
MEP-1	SITE PLAN: MEP
P-2	PLUMBING FLOOR PLAN
P-3	RISER DIAGRAMS, DETAILS AND SCHEDULES
P-4	DETAILS
M-2	HVAC FLOOR PLAN
M-3	MECHANICAL SCHEDULES
E-2	LIGHTING FLOOR PLAN
E-3	POWER FLOOR PLAN
E-4	ELEC. SCHEDULES AND DIAGRAMS

INDEX OF DRAWINGS

COVER SHEET
A0.10	INDEX, CODE INFORMATION, CONVENTIONS
A0.21	EXITING PLAN

CIVIL

C1	SITE PLAN
C2	PAVING & GRADING PLAN
C3	STORM SEWER PLAN
C4	EROSION CONTROL NOTES
C5	SITE DETAILS

ARCHITECTURAL

D2.01	SITE DEMOLITION AND BUILDING PHOTOGRAPHS
D2.02	DEMOLITIOIN FLOOR PLAN AREA 'C', 'D', 'E' AND 'F'
A2.00	COMPOSITE FLOOR PLAN
A2.01	SCHEDULES, CASEWORK, DOOR WINDOW TYPES
A2.02	FLOOR PLAN AREA 'A' AND 'B'
A3.00	COMPOSITE REFERENCE ROOF PLAN
A3.01	ROOF DETAILS
A4.00	TYPICAL MOUNTING HEIGHTS
A5.00	PLAN DETAILS
A6.00	EXTERIOR ELEVATIONS
A7.00	WALL SECTIONS AND DETAILS
A10.02	REFLECTED CEILING PLAN AREA 'A' AND 'B'

STRUCTURAL

S1.01	GENERAL NOTES
S1.02	GENERAL NOTES
S2.01	FLOOR FRAMING PLANS
S2.02	ROOF FRAMING PLANS
S3.01	CONCRETE SECTIONS AND DETAILS
S3.02	CONCRETE SECTIONS AND DETAILS
S4.01	STEEL SECTIONS AND DETAILS
S4.02	STEEL SECTIONS AND DETAILS

MECHANICAL

M1.02	HVAC MECHANICAL PLAN AREA 'A' AND 'B'
M2.02	PIPING MECHANICAL PLAN AREA 'A' AND 'B'
M3.02	MECHANICAL DETAILS
M4.02	MECHANICAL SCHEDULES

ELECTRICAL

E2.02	ELECTRICAL POWER FLOOR PLAN AREA 'A' AND 'B'
E3.02	ELECTRICAL LIGHTING FLOOR PLAN AREA 'A' AND 'B'
E5.01	ELECTRICAL DETAILS AND SCHEDULES
E5.02	ELECTRICAL DETAILS AND SCHEDULES
E6.01	ELECTRICAL RISER DIAGRAM AND SCHEDULES

PLUMBING

P2.02	PLUMBING FLOOR PLAN AREA 'A' AND 'B'
P4.01	PLUMBING DETAILS

TECHNOLOGY

TS3.02	FIRE ALARM TECHNOLOGY FLOOR PLAN

FIGURE 3.1 Sheet Indexes

Types of Drawings

Within each category, the sheets are drawn to be a specific type, or kind, of drawing to convey the various types of information required to build the project. There are five primary types of drawings found in each category, plan views, elevations, section, details, schedules, and schematics. Each type of drawing is designed to convey certain kinds of information using a specific format for information presentation. We will look at both the kind of information and information presentation aspects as we study the types of drawings.

To fully understand the types of drawings, we need to understand how the architect, engineer, and drafters view objects. Referencing Figure 3.2(A), note the object in the box.

The first thing the drawing creator considers is which surface or element constitutes the front of the object. This is key to understanding the position of the other views of the object and how they relate to each other. For example, in our illustration, the draftsman has indicated the front of the object—in this case, a building—with an arrow. He has designated this as the front, and we should try to mentally position ourselves as though we were standing on that side of the building looking directly at the front. This is called *establishing our line of sight*. Once line of sight is established with the front of an object, we can then identify the other side or views of the object.

If viewers who are orienting themselves with the line of sight to the front of the object mentally reached out their right hand, they could touch the side of the object on that side—this would be considered the right side of the object. From the same orientation, if viewers reached out their left hand and touched that side of the object, that would be considered the left side. The same would be done from the front line of sight to establish the top, rear, and bottom views. Each view of an object is labeled or identified by its relationship to the front view of the object.

The next issue that the creators of the drawings consider are the planes relative to the various lines of sight that will be used to show the object. This type of multiview drafting technique is called *orthographic projection*. In Figure 3.2(B), note that the sides of the box are now considered geometrical planes, or surfaces that we can project the line of sight images of the object onto for viewing. After the line-of-sight view of the object is projected out, these planes can be separated (Figure 3.2[C]) for individual viewing.

Users of the working drawings must be able to look at the front view of an object and the right view of an object and understand where the corners meet, as shown in Figure 3.2(D). This is very important for individuals who use the drawings to understand, because the separated views of the object will probably be on different sheets, and users will have to comprehend where their line of sight is as they move from view to view and sheet to sheet. Later in this chapter, we will look at some symbols that will aid in accomplishing this.

Once comfortable with this orientation concept and line of sight, we can now look at the types of drawings, since they are derived from the box concept in the example.

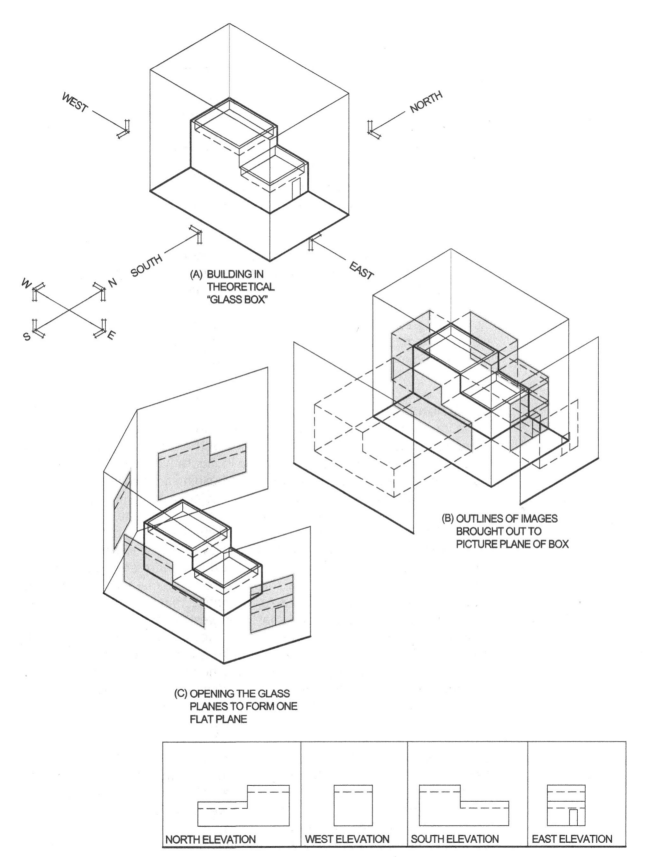

(A) BUILDING IN
THEORETICAL
"GLASS BOX"

(B) OUTLINES OF IMAGES
BROUGHT OUT TO
PICTURE PLANE OF BOX

(C) OPENING THE GLASS
PLANES TO FORM ONE
FLAT PLANE

| NORTH ELEVATION | WEST ELEVATION | SOUTH ELEVATION | EAST ELEVATION |

(D) THE ORTHOGRAPHIC VIEWS
ON A SINGLE PLANE

FIGURE 3.2 Building Elevation Study

PLAN VIEWS

Probably the most important and informative type of drawing is the plan view. This is a view of the object as though you were suspended above it, with your line of sight looking straight down at the object. It is also considered the top view of the object, as derived in Figure 3.3. The information shown on plan views is two-dimensional—length and width—giving us the basic *size* element of the building. There are many plan views in a working drawing package, some of which look at the building from the outside and some of which look at the building from the inside. For example, there will be plan views of the construction site, foundation, structural members, and roof, all from outside the building. The other type of plan view we see is called a floor plan, in architectural drawing terms, because it views the building from inside and shows more detail on the floor layout, or *location* of rooms, corridors, door, windows, and so on. Figure 3.3 shows how this view is derived from the top view. An imaginary cutting plane is used to cut the building in half horizontally. The top half is removed, and we look inside at the building details from the same line of sight as the top view. As we progress through the various categories in the working drawings, we will find many different plan views and floor plans. Figure 3.4 shows a finished commercial building plan view.

ELEVATIONS

Elevations are used to view the third dimension of an object. They give us the depth or height dimension and can be of the exterior or interior of the object—a building, in our case. As shown in Figure 3.2(C) and (D), the front, right, and left sides, and rear views of the building are shown as our line of sight progresses around the building, thus giving us more information such as the *shape* of exterior building details. As we see these views, we now need to begin to think about them in architectural drawing terms. The front view in those terms is now called the front elevation. The right, left, and rear views now are called *elevations*. Although this is more descriptive for the construction industry, there is a still more accurate method for describing elevations. Looking at Figure 3.2(A), we notice that there is a directional arrow that indicates north, south, east, and west. In looking at this compass indicator, we can see that the front view, or elevation, is facing to the south. Therefore, the draftsman will probably label the elevation for the front of the building, south elevation. This is the method most often used to indicate elevations in the working drawings for a commercial construction project. Figure 3.5 shows exterior elevations of a commercial building.

SECTIONS

Sections are views of the building used to show material placement and construction detail. These views will be established using cutting planes like the one shown in Figure 3.3, where we cut away the top half of the building to show the interior details of the floor plan. Drafters use these cutting planes on various parts of the building at different angles on the floor plans and elevations to expose hidden detail of how materials were placed to construct walls, floors, structural members, and other aspects of the project that you could not see through the other views. Sections will

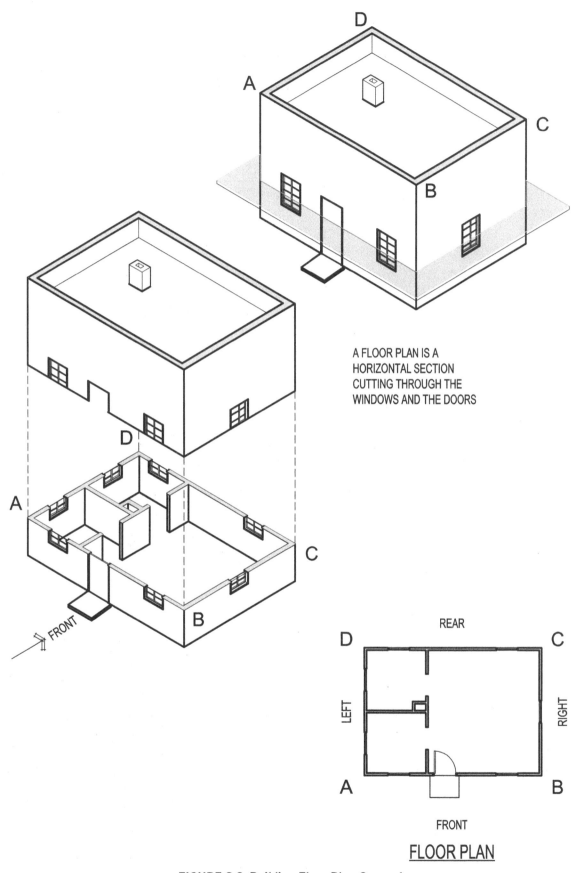

A FLOOR PLAN IS A
HORIZONTAL SECTION
CUTTING THROUGH THE
WINDOWS AND THE DOORS

FLOOR PLAN

FIGURE 3.3 Building Floor Plan Concept

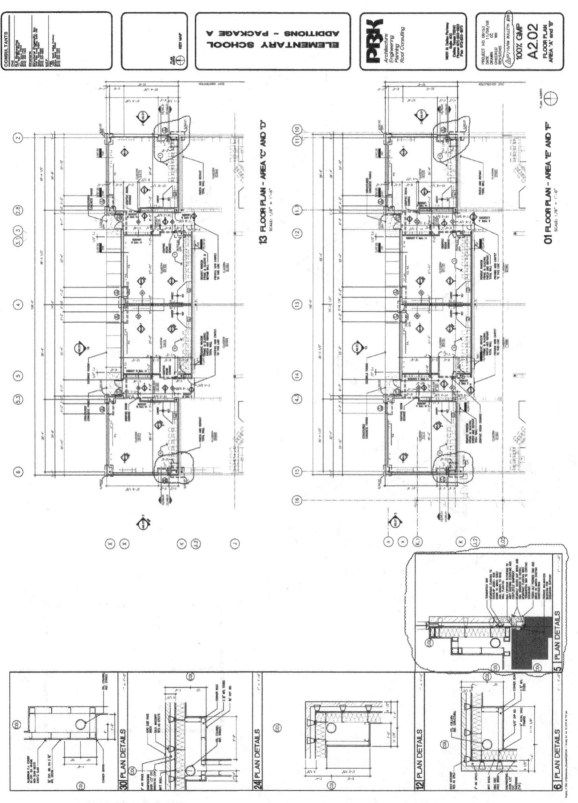

FIGURE 3.4 Plan Views

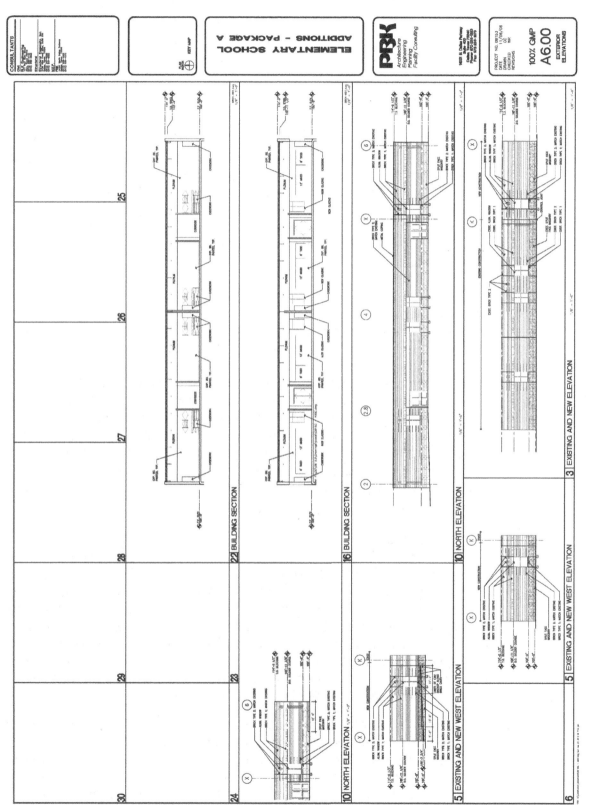

FIGURE 3.5 Elevations

be viewed from a specific line of sight that will be indicated on the drawing at the specific location, as indicated on the drawing by the cutting plane. Later in this chapter we will look at symbols used to indicate the line of sight and cutting plane. Figure 3.6 shows sections taken through walls of a commercial building.

DETAILS

Even though the section views show a significant amount of information on the construction of the building, they are drawn at a scale that may not allow scrutiny of critical locations of the building such as where the roof intersects the exterior walls, or where wall construction attaches to structural columns. Details are a portion of a section view drawn at a larger scale to show more of the details that cannot be seen at the smaller scale. The line of sight for the detail will be exactly the same as the original view from which it was derived. Figure 3.7 shows details for a commercial building.

SCHEDULES

There are certain elements used to construct a building that would require a lot of definition on the drawings to understand dimensionally. To avoid cluttering up the drawing field, the drafter will use schedules to catalog the information. Symbols, which we study later in this chapter, will be used to delineate these individual elements and will refer back to the particular chart, or schedule, that contains the information for that element. Elements that typically will be described in a schedule would include landscaping, doors, windows, structural piers, columns and beams, mechanical equipment, and electrical breaker panels. Figure 3.8 shows a door and room finish schedule for a commercial building.

SCHEMATICS

Schematics are used in the working drawings to show how a system or complex idea works, but do not show the exact locations where the components are found. Examples of schematics in the working drawings include the One-line Diagram, Figure 9.2, found on Page 173, and the Plumbing Riser Diagram, Figure 10.7, found on Page 194, in this text book. The schematics allow conduits and piping to be shown form fixture to fixture, the sequencing of fixtures, and the controls used with them and how they are connected. However, schematics do not show the locations where the fixture and control components are located.

These six types of drawings allow the architect, engineer, and draftsman to describe the elements of size, shape, and location of the components of the building in an efficient and progressive manner. You will find plan views, elevations, sections, details, schedules, and schematics in each of the categories of the working drawings: civil, architectural, structural, mechanical, electrical, and plumbing. Now we must study what symbols are used to move from sheet to sheet in the drawings and how they are used.

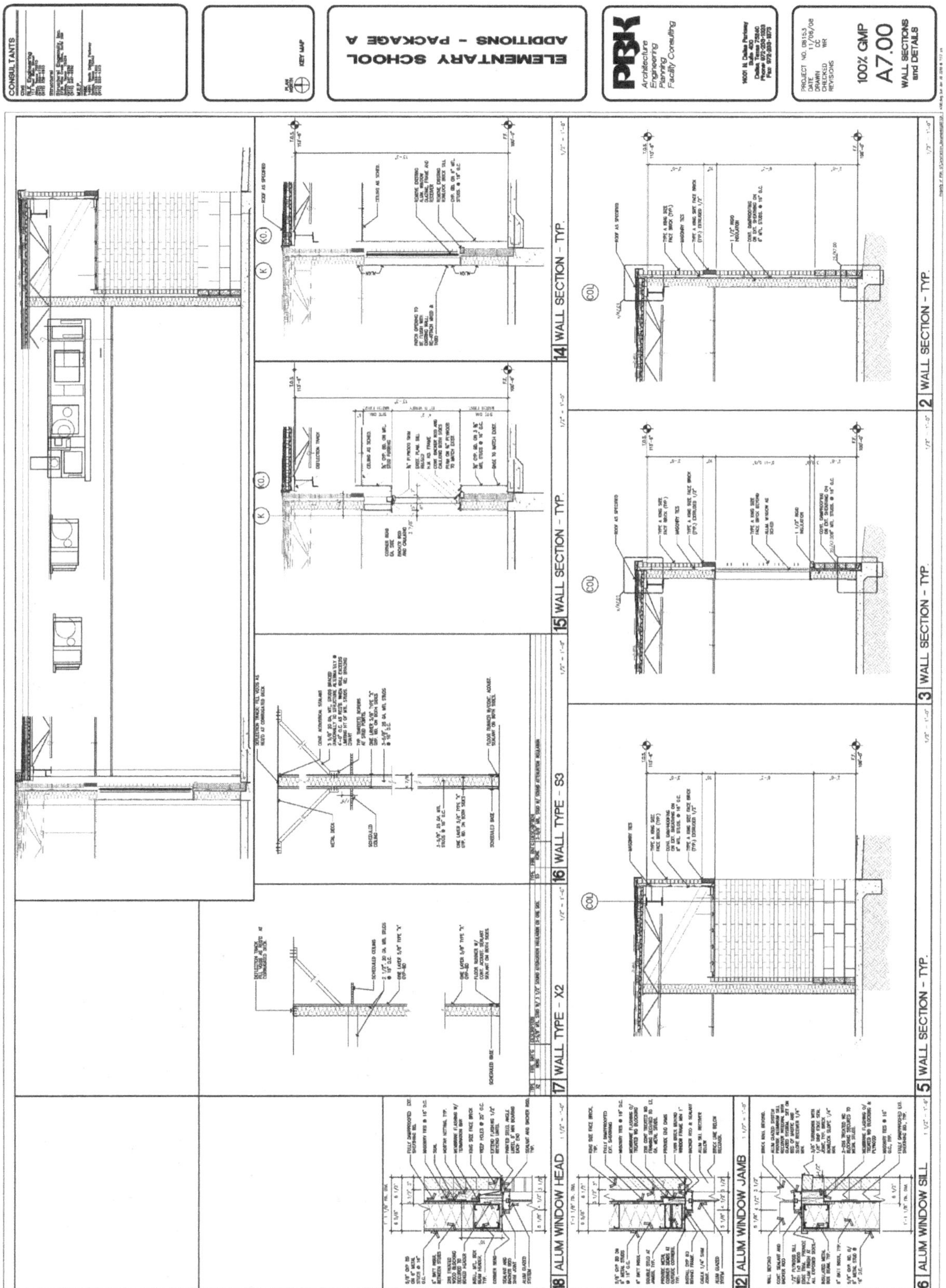

FIGURE 3.6 Sections

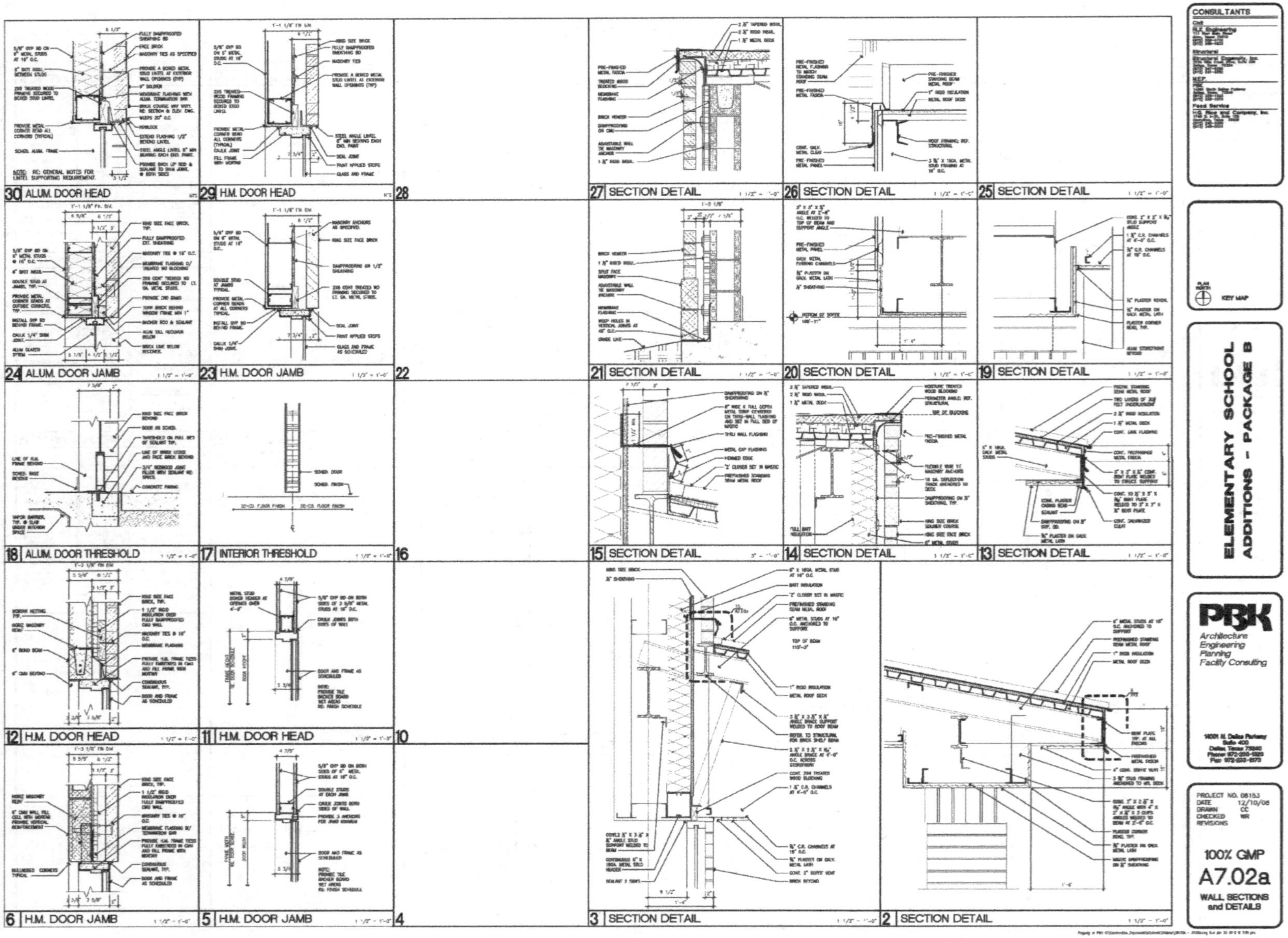

FIGURE 3.7 Section Details

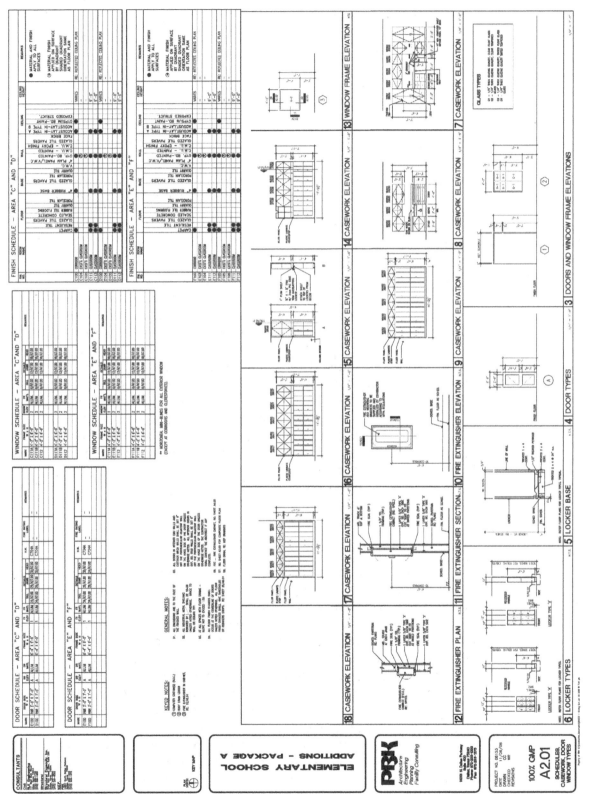

FIGURE 3.8 Schedules

THE DRAWING ELEMENTS

The drawing elements are composed of the standard symbols that are used on the drawings to direct us to different sheets for building information and are used to convey what is on the drawings and contain lines that have specific meanings. Let's look at the symbols first.

Symbols Used in Creating the Working Drawings

There are many types of symbols used in creating the working drawings. The primary reasons for symbol use are to minimize the amount of time that would be required to draw the actual materials used and to allow the drawing users to locate and identify information quickly. By using symbols, we can simplify details and accelerate the creation of the drawings.

MATERIAL SYMBOLS

When we think of the symbols used in the drawings, we typically think about those symbols that represent materials. There are as many symbols for this as there are materials—more than can be effectively learned in this course. It is not feasible to know all the symbols, but it is important to understand how to identify what the symbols represent in the set of drawings one is working on. Most architects, engineers, and drafters will create a legend on each sheet indicating the symbols used on the sheet and which materials they represent. Most of these symbols are developed by the American National Standards Institute (ANSI) and the American Institute of Architects (AIA). These symbols are not scalable and are applied to indicate typical placement, unless their location is dimensioned. Therefore, we will not be spending a significant amount of time with these in this chapter. We will instead be looking at specific material symbols when we study the contents in each category of the working drawings in the chapters in Section II and will concentrate on callout symbols in this chapter.

CALLOUT SYMBOLS

These are the symbols that will assist us in moving from view to view and sheet to sheet to locate information. They provide the directions for moving between the six different types of drawings. There is no standard appearance to these symbols, but there are elements to these symbols, which are consistent from architect to architect. The drawing creator will provide us with a legend indicating what the symbols are that are used in each drawing package. The first type of drawing we learned of is the plan view, and most of the callout symbols can be found on those types of drawings, although they are on the other types of drawings as well. That is why, as we will see later, we start with the plan views when beginning to work with the drawings. Something to note: When you pick a plan view in a category in the working drawings, the callouts on that plan view will not reference elevations, sections, or details in another category. For example, if you start looking on a plan view in the mechanical category of the drawings, all of the callouts on the plan view will reference views in the mechanical category, not the architectural, structural, electrical, or civil categories.

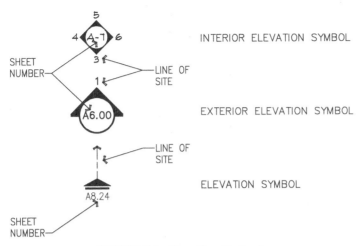

FIGURE 3.9 Elevation Callouts

The first type of callout symbol we want to look at is the elevation callout. This callout will typically be placed on the plan view to indicate to the reader that there is a view of the building, either interior or exterior, showing a vertical surface of the building. Look at Figure 3.9 and we will see some examples of this symbol. Although the symbols in Figure 3.9 vary in actual appearance, all elevation callouts will convey three pieces of information: line of sight, the sheet number on which the elevation can be found, and the identification number of the elevation. Even though the Figures in 3.9 appear different, the elevation callouts should both convey the same information. We have discussed line of sight before, and the same principles apply here: It is the direction the viewer would be looking if he were positioned facing the building as indicated. The only other information we need is to know what sheet in the drawing package to look on and what the identification number is for the elevation we want to view. Exterior elevation callouts will typically be shown on the plan view outside the building perimeter, and interior elevation callouts will be shown inside the building perimeter. All elevation callouts will show the three basic pieces of information, line of sight, sheet number, and identification number.

The next type of callout symbol we need to consider is the *section view* callout. This symbol has the elements shown for the elevation callout, plus one more piece of information, the cutting plane line. Recall our discussion of plan views and how we obtain a floor plan view of the inside of the building, as shown in Figure 3.3; the same cutting plane principles apply here. Figure 3.10 shows samples of section view callouts. Notice that each callout provides the line of sight, the sheet number the section view is located on, and the identification number of the section view, just as in the elevation callout. In addition to these elements, there is a line, called a cutting plane, drawn through the building, or some portion of the building, indicating that when the viewer goes to the sheet indicated and views the section, he will be looking at the building as though he were looking at a cut through all the materials at the location where the cutting plane is drawn through the building. The line of sight will indicate the orientation of the viewer to the cutting plane, or the direction he will be looking when he gets to the section view. Even though the section view callouts will vary

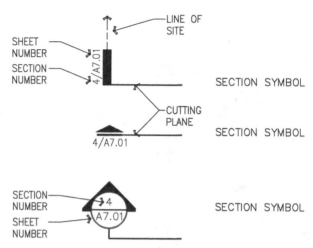

FIGURE 3.10 Section View Callouts

in appearance, they will all convey four pieces of information, line of sight, the cutting plane, the sheet the section view can be found on, and the identification number of the section view.

When we discussed how detail types of drawings are derived, we indicated that they originate from a section view and that the line of sight does not change. In addition, we noted that the detail view shows only information from the section view, but at a larger scale so that we can see more construction detail. The location that the detail view comes from on the section view, and how much of the section shown in the detail view is indicated by a dashed line that surrounds all the information from the section view that will become the detail view. This dashed line will also have a leader line that will show the indicators for the sheet number the detail view is on and the identification of the detail view, see Figure 3.11.

As noted in our description of schedules, we want to refer the viewer back to a schedule to see descriptive information on certain building elements to avoid having all that data consume the plan views or other views. To accomplish this, we will use geographic symbols such as squares, hexagons, circles, ellipses, and rectangles. These schedule callouts will be placed next to the object indicated with either a number or letter to show where on the schedule the information is listed. The drafter will chose which symbols he will use for each element listed in a schedule. Look at Figure 3.12. The drafter has chosen to use circles with a letter inside for

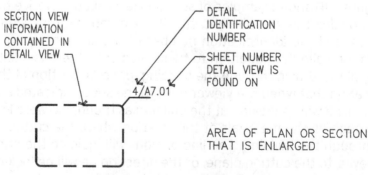

FIGURE 3.11 Detail Callouts

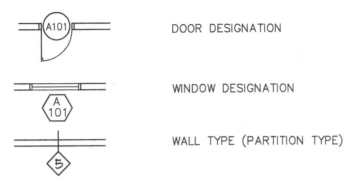

DOOR DESIGNATION

WINDOW DESIGNATION

WALL TYPE (PARTITION TYPE)

FIGURE 3.12 Schedule Callouts

windows and a square with a number inside for doors. Each door that has the same elements will have the same callout on the schedule, a square with a number inside. For example, if 20 doors are exactly the same as door 1, they will probably all have the same schedule callout symbol. However, if there are 50 doors on a project and the drafter wants to give them each an individual schedule callout, he can. The drawing reader must note which symbols are used for which building element and find the appropriate schedule and number to view data for that element.

Again, none of the callout symbols will cross drawing categories. If you start with a callout in one category, all the drawings types referenced will be in that category. The drawing reader will have to look at the legend provided by the drafter to understand exactly which styles of callout symbols are used on the working drawings he is looking at.

The Drawing Sheet Format

Although each drawing contains information for different pieces or elements of the building, most drafters follow a basic format for their sheets so that information is consistently located on the sheets. Look at Figure 3.13. This is a typical format for drawing sheets in the working drawings.

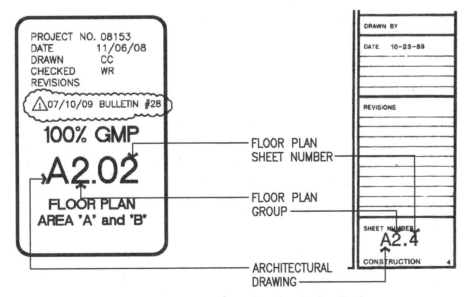

FIGURE 3.13 The Block/Drawing Numbering Systems

TITLE BLOCKS

Because we read from left to right, drawing packages are fastened on the left side like a book. This side contains little or no information. On the lower-right side corner, or along that side of the drawing sheet, the drawing *title block* is shown. This location allows us to page through quickly without having to fully open each sheet to find the one we are looking for. The title block information will typically contain information such as the architect's name, project title and number, sheet number and category, sheet title and release date, revision box, and any architectural and engineering seals.

SHEET NUMBERING SYSTEMS

Note the sheet numbering systems in Figure 3.13. There are various systems used by architects; however, they will always indicate in the legend which system is used in the drawing package you are looking at. As a minimum, the sheet numbering system will indicate the category the sheet belongs to:

C–Civil

A–Architectural

S–Structural

M–Mechanical

E–Electrical

P–Plumbing

The sheet number will also indicate what number the sheet is in some kind of sequential order.

NOTES

As a general rule, notes will be positioned in the upper-left corner of the drawing sheet, and, depending on the number of notes required for the information drawn on the sheet, may stretch all the way across the top of the sheet. The notes will usually pertain only to the sheet they are on. General notes that pertain to all the sheets in a category will typically be on the first sheet in that category.

NORTH ARROWS

North will be indicated on all plan views as discussed previously. True north will be given, and used, if the project is not at some difficult angle for calculations or reference. If the project as at some difficult angle to north, a plan north will be established and indicated, as in Figure 3.14 for ease of reference. As with maps, the drafter will typically orient north as the top of the drawing, which is preferred (north being to the left is the second option).

DRAWING FIELD

The center of the sheet is typically referred to as the drawing field and contains the graphical information on the project, as indicated in Figure 3.15.

TRUE
NORTH

PLAN
NORTH

01 FLOOR PLAN

SCALE: 1/8" = 1'-0"

FIGURE 3.14 North Arrows

There are several drawing elements to note in this area. The object being drawn—in our case, a building—is drawn using certain line conventions to indicate surfaces and components of the object. Some of the lines are used to indicate reference to other portions of the object or building. In Figure 3.16, there are several types of lines used to indicate various elements of the drawing. Following are the basic definitions of the lines used:

- *Object line.* A medium-width, solid line used to indicate surfaces and features of the building that are visible from an indicated line of sight.

- *Hidden line.* A thin-width, dashed line used to indicate surfaces and features of the building that are not visible from an indicated line of sight.

- *Center line.* A thin-width line with alternating short and long dashes that are used to indicate the center of an object or building elements that are arranged along a continual line.

- *Leader line.* A thin-width solid line used to direct the drawing viewer from a note to the specific location on the drawing field being referenced.

- *Break line.* A thin-width line used to indicate that the building continues beyond what is shown but for clarity was stopped at the break line.

- *Dimension line.* A thin-width, solid line used to indicate the distance over which a dimension applies, typically ending with arrowheads at both ends to indicate the dimension limits. It applied parallel to the surface or features being dimensioned. Dimensions in construction are shown in feet and inches and dimension lines are continuous with the dimension placed in the center just above the line. As in Figure 3.16, dimensions always give actual sizes of building elements regardless of the scale the drawing is drawn to.

- *Extension lines.* A thin-width, solid line used to extend a surface feature or element of the building to allow clear dimensioning; the extension line does not contact the object being drawn.

- *Section lines.* Thin-width lines used to indicate where the object has been removed using a cutting plane to expose hidden construction details, sometimes used in various patterns to indicate materials.

In Figure 3.16, note the elements of the section callout used. Also, along the right side of the plan view from top to bottom, you see an "A" and "B" in hexagons. Along the top and bottom of the plan view you see a "4" in a hexagon. These are indicators for a coordinate grid system.

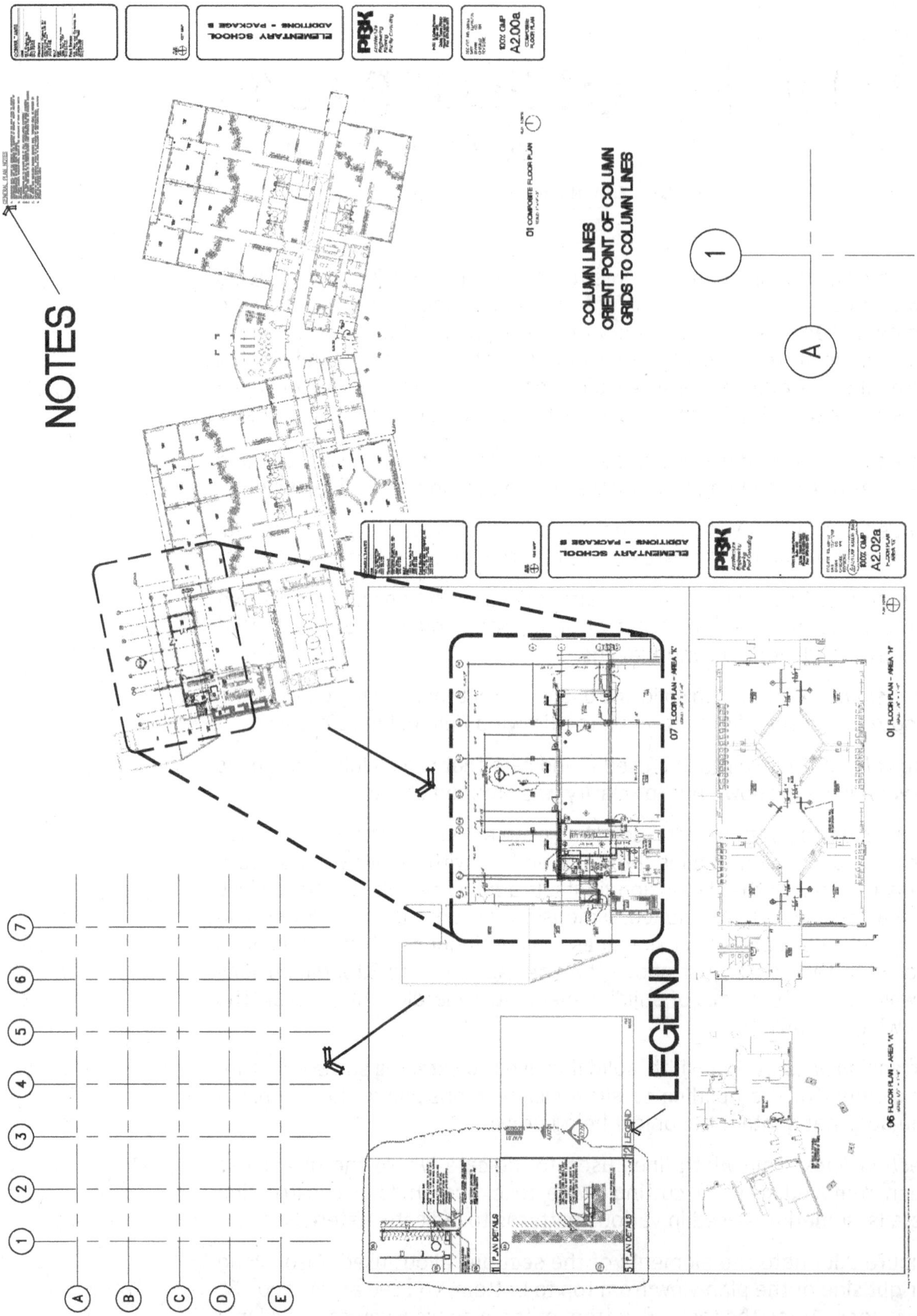

FIGURE 3.15 Drawing Field Concepts

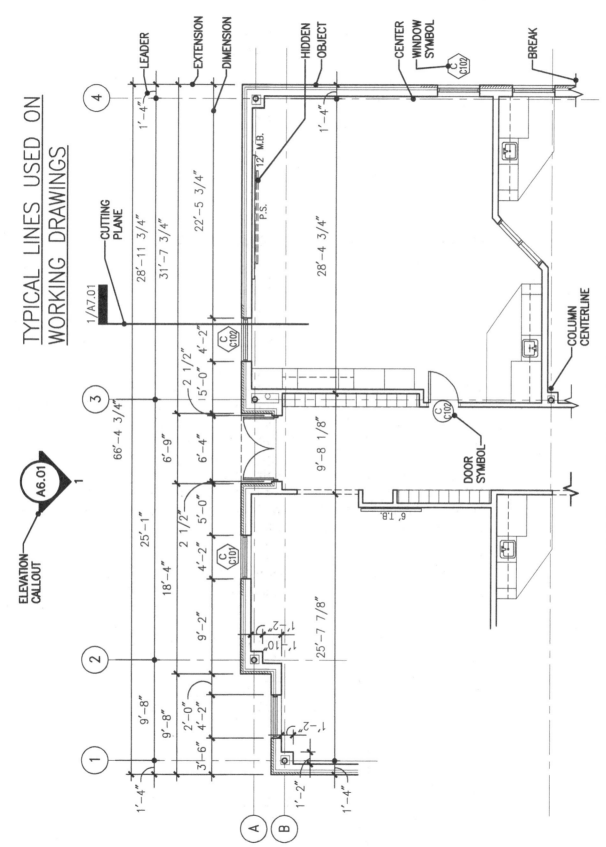

FIGURE 3.16 Types of Working Drawing Line Conventions

51

COORDINATE GRID SYSTEM

This grid system, technically a Cartesian coordinate system, is established typically along column lines of the building to provide easy reference to various locations on the drawing. Look at Figure 3.15. This grid system usually consists of very thin lines going both directions on the drawing field with numbers used in one direction—say, top to bottom—to identify the horizontal lines, and numbers going the opposite direction to identify the vertical lines. It typically is laid out to intersect with the structural columns or some other structural member. This grid system allows individuals in multiple locations to be able to reference the same point on a drawing when discussing the project. This grid system will remain the same throughout all of the categories in the working drawings.

DRAWING SCALES AND DIMENSIONS

Dimensions are depicted on construction drawings as shown in Figure 3.17 with extension lines drawn from the object and straight continuous lines with either arrowheads or architectural tick marks. The feet and inches are always shown. If there are no feet in the dimension, then "0" is used as a place holder. Dimensions should not be confused with scales.

As we progressed through the various types of drawings, plan views, elevations, sections, details, and schedules, it might have became evident that as each type provided different pieces and pictures of the building, each would present this information drawn at different scales. The use of scales allows the drafter to draw an accurate representation of an object smaller or larger as necessary to conveniently convey the information for the viewer. A plan view and elevations might be drawn at a scale of 1/8 inch = 1 foot; sections drawn at 1/4 inch = 1 foot, and details drawn at a scale of 3 inches = 1 foot. Each one is drawn at a progressively larger scale to allow more detail to be presented.

Scales are simply ratios used to compare the original of some object to a smaller version. 1:1 ratio indicates full scale. We are stating that 1 foot = 1 foot in that ratio. A 0.5:1 ratio indicates half scale. We are stating that

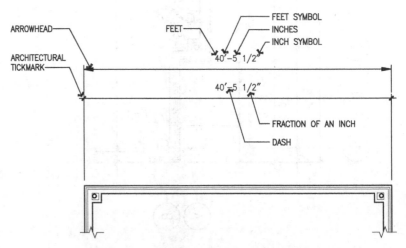

FIGURE 3.17 Method of Showing Dimensions

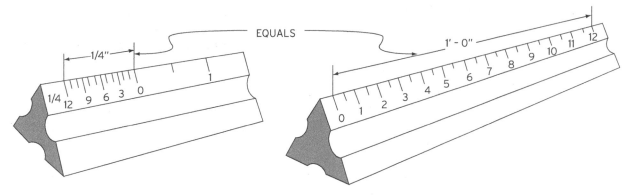

FIGURE 3.18 Arch Scales

1/2 inch = 1 inch, which can be extrapolated out to any ratio: 6 inches on the drawing equals 12 inches in the construction.

In architecture, scales, or ratios, are compared to 1 foot. In Figure 3.18, the ratio is 1/4 inch = 1 foot, represented as 1/4″ = 1′-0″. With the understanding that 1/4 inch is smaller than 1/2 inch, we can know that anything drawn at a scale of 1/4″ = 1′-0″ will be smaller than if drawn at 1/2″ = 1′-0″. Now we can relate to the different scales used on the different types of drawings.

The scale used to create the working drawings is called an architect's scale. This scale, typically a triangular-shaped ruler, has one full scale of 12 inches divided into sixteenths of an inch, and ten other scales with increments from 3/32 inch to 3 inches. The scale increments of 1/8″, 1/4″, 1/2″, and so on are all used to represent 1′-0″. Look at Figure 3.19. In this example, we can see the dimension 5′-6″ and how long it actually is when drawn at the different scales. Notice that each scale begins at one end and reads to the opposite end with multiple whole feet shown moving toward the center of the scale, and one whole foot, 12 inches reading from zero to the end of the scale. Each side of the scale has two scales. The scale has a "0" as its starting point. From zero, you would read to the center of the scale for the whole foot increments. You would read from zero to the end of the scale to determine inches. When reading a specific scale, one must now begin to think of counting spaces in the scale increment used to indicate feet. For example, if I am reading the 1/4″ = 1′-0″ scale, I must think that each 1/4″ space I measure equals 1′-0″. Each scale has one whole foot, or 12 inches, indicated next to zero. This allows us to understand that if some line we are measuring extends less that one whole foot at the scale

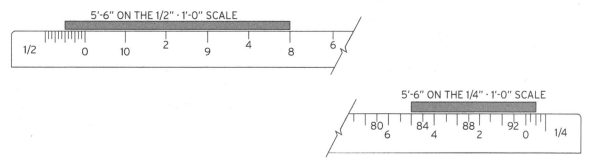

FIGURE 3.19 Dimensions Scales

we are measuring, we can determine the inches in length that line' is. This is fairly simple because if we are working on the 1/4"=1'-0" scale, we know that a 1/4" space has 12 inches in it. Reading from zero, we can determine that half the distance of one is 6". One quarter of the distance is 3".

Looking at a scale, you find the 1/8" and 1/4" scales. They are on the same side of the architect's scale. The 1/4" scale reads from right to left and the 1/8" scale reads from left to right on the scale. If I want to measure a line that is 5'-6" long at a scale of 1/4" = 1'-0", I will start reading the 1/4" scale, moving from zero to the left five whole 1/4" spaces. Since on the 1/4" scale, one space equals 1 foot, we have moved 5 feet. Then, moving to the right another half of a 1/4" space, we will have moved 6 inches. By putting one end of a line on a whole increment of the scale being used and allowing the other end to lap past zero and into the inches measurement, we can quickly determine the length of the line being measured.

These scales are used to draw the building elements at the right ratios for conveying the construction information to the viewer. Using these scales, we can understand the sizes of the building elements as they relate to each other.

Another critical aspect of scales is that they are only accurate on the original drawings. Expansion and contraction due to changes in temperatures or moisture in the air affect how long or short distances are on the drawings, as the paper reacts to the atmosphere. This has a less of an impact at a scale of 3" = 1'-0" than at 1/8" = 1"-0" but the consequences can be catastrophic for either scale, as the wrong amount of materials can be purchased or excessive labor spent to dig a trench too long. As a result, unknown dimensions must be calculated from known dimensions. Calculating dimensions from known dimensions will give the user the correct distances. Scaling any drawing will result in the wrong measurement being used.

To obtain accurate dimensions you should use given dimensions on the drawing to calculate these dimensions that are omitted.

In Figure 3.20 we can see how this is accomplished. Notice dimension A, from the window to the corner of the building. If this dimension is needed, instead of scaling the copy of the original drawing, we should use the existing dimensions, 18'-4" and subtract 4'-2" and 9'-2", to obtain the distance. If we scale the copy of the original, our distance will be incorrect. This could have a dramatic influence on our ability to provide a correct estimate of materials and cost for completing the project, or cost us valuable construction time in the field if we have to rework the corner because of an incorrect dimension.

This is the only acceptable way to obtain distances on copies of the working drawings. Do not scale copies of the working drawings to obtain dimensions. The typical plan dimensioning conventions are shown in Figure 3.21 and illustrate some of the techniques used in dimensioning construction drawings.

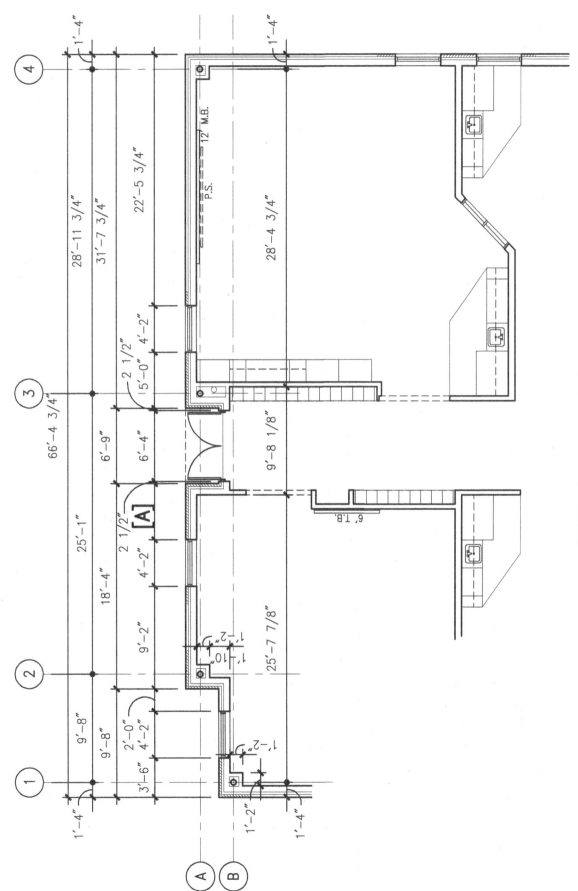

FIGURE 3.20 Arch Dimensioning

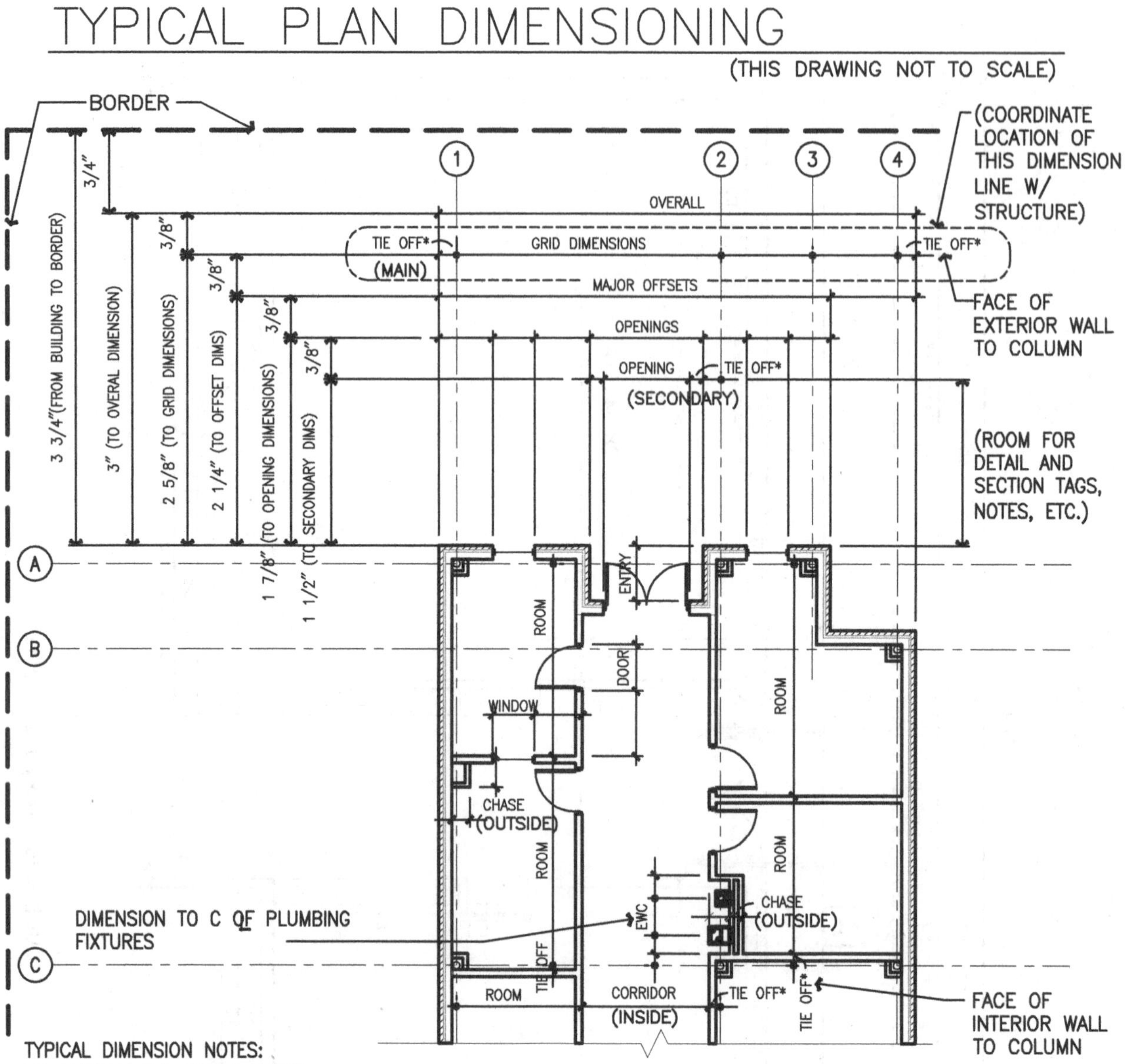

TYPICAL PLAN DIMENSIONING

(THIS DRAWING NOT TO SCALE)

TYPICAL DIMENSION NOTES:

1. TIE OFF INTERIOR & EXTERIOR DIMENSION STRINGS TO A STRUCTURAL GRID OR TO A WALL THAT IS OBVIOUSLY TIED TO A GRID.
2. EXTERIOR DIMENSIONS: DIMENSION TO OUTSIDE OF WALL FINISHES.
3. INTERIOR DIMENSIONS: DIMENSION TO TOP SIDE AND RIGHT SIDE OF WALLS
4. CORRIDORS, STAIRS, AND ELEVATOR SHAFTS: DIMENSION TO INSIDE FACE OF WALLS
5. ENLARGED TOILET PLANS:
    5.1. DIMENSION TO INSIDE FACE OF WALLS
    5.2. INCLUDE CRITICAL TAS/ ADA CLEARANCES
    5.3. DIMENSION TO CENTERLINES OF PLUMBING FIXTURES
6. INTERIOR DOOR OPENINGS:
    6.1. DIMENSION TO SIDE OF FRAME CLOSEST TO ADJOINING PLAN ELEMENTS.
    6.2. DIMENSION BOTH SIDES OF FRAME WHERE FRAME OPENINGS ARE GREATER THAN 6'–4".
7. EXTERIOR DOOR OPENINGS: DIMENSION BOTH SIDES OF FRAME, TYPICAL.
8. INTERIOR AND EXTERIOR WINDOW OPENINGS: DIMENSION BOTH SIDES OF FRAME, TYPICAL.
9. SHAFT, CHASE, AND OTHER BUMP–OUTS: DIMENSION TO OUTSIDE OF WALLS.

FIGURE 3.21 Arch Dimensions

SUMMARY

In this chapter we have explored the working drawings structure. We have learned about the six categories the working drawings are grouped into: civil, architectural, structural, mechanical, electrical, and plumbing. We have also studied the five types of drawings found in each category, plan views, elevations, sections, details, and schedules, and identified what kinds of information can be found on each type of drawing.

We have studied the various elements found on drawings, including callout symbols, the drawing sheet format, north arrows, drawing line conventions, and the grid system. Finally, we have studied scales and how they are used by the drawing creators to convey accurate detailed information on the different types of drawings and that the copies of the working drawings should not be scaled. Dimensions should only be calculated from given dimensions.

One other important thing we learned in this chapter is that the drawings are created to show us the size, shape, and location of building elements. This will be an important part of our process for finding information quickly in the contract documents, as we will see in a later chapter.

▓ CHAPTER 3 LEARNING OUTCOMES

- The information in the working drawings shows us size, shape, and location of elements of the project.

- The sheets in the working drawings are grouped into six primary categories: civil, architectural, structural, mechanical, electrical, and plumbing.

- There are five primary types of drawings in each category of the working drawings: plan views, elevations, sections, details, and schedules.

- Plan views have a line of sight as though the viewer were positioned directly above the object on the drawing.

- Elevations typically are referenced by the direction—north, south, east, or west—that they face.

- Section views use a cutting plane to indicate material placement and construction details inside walls and either building elements at a specific location.

- Detail views are areas from a section view drawn at a larger scale.

- Schedules are chartlike drawings used to organize information that would clutter up the other drawing information if placed on the same views.

- A section view callout contains four elements: a cutting plane, a line-of-sight indicator, the identification number of the section, and the sheet number it can be found on.

- Various types of lines are used to indicate different building elements on the drawings.

■ CHAPTER 3 QUESTIONS

1. The drawings show which elements used to interpret the construction information they contain?

2. List the six primary categories sheets are grouped into in the drawings.

3. What are the five primary types of drawings in each category?

4. What is the line of sight in a plan view?

5. How are elevations referenced in the drawings?

6. How are detail drawings derived?

7. Describe schedule drawings and what they are used for.

8. Describe what a section view is and the elements it consists of.

4

SPECIFICATIONS

OVERVIEW

As stated in Chapter 2, the specifications are the primary written contract document for a commercial construction project. Typically, they describe the types and qualities of materials used and the different methods of construction employed. They will also provide some dimensions—mainly those that would be too small and difficult to find if they were included in the working drawings. They are significant contract documents in that contradictions between the other contract documents are frequently resolved through the information in the specifications. In the event of conflicts between the information presented on the drawings and in the specifications, the information in the specifications will typically govern. The specifications are developed by the architect and his consulting engineers, or sometimes spec writers, who have detailed knowledge of construction methods, materials, equipment, and building codes. The information provided to us about the project will pertain to quality, materials, and methods. In this chapter we will learn more about the structure of this document and the function of the specifications, and how the information contained within is used. We will also look at the various contract documents included in the specifications.

SPECIFICATION USERS

The specifications are written for use by many of the team members involved in the construction process. Each member will look at them from his or her own perspective to satisfy construction requirements.

The general contractor is interested in the specifications since he will be contractually committing to provide all the materials listed. The general

Print and Specifications Reading for Construction, Updated Edition. Ron Russell.
© 2024 John Wiley & Sons, Inc. Published 2024 by John Wiley & Sons, Inc.
Companion website: www.wiley.com/go/printspecreadingupdatededition

contractor will also be interested in what construction requirements will be needed to put the materials in place. This is important because the contractor will be bidding on this information and then building from it if awarded the project. This same interest holds for to the subcontractors who might be used to accomplish any of the work. The general contractor is contractually committed to the owner for the entire project and must ensure that all subcontractors meet the specifications' requirements.

The owner is interested in the specifications since they represent the requirements for the construction materials and placement methods necessary to provide the facility to serve his or her needs once the building is finished. Specifications also represent the level of quality that is expected from the general contractor. The owner's project manager will also use the specifications to monitor and verify that the materials and methods being specified are being provided and used.

Various estimators will use the specifications to develop costs. The owner's or architect's estimators will use the specifications to determine budgetary costs for use in determining the quality the owner can afford or that is necessary for the function of the building. The general contractor's estimators will use the specifications to determine what costs he will incur to provide what is required. It is a competitive advantage for contractors when materials and products are specified that they can obtain special pricing for. A contractor that purchases 75 percent more of a particular item than other contractors is likely to obtain a more favorable price and have a competitive advantage over the other bidding contractors in being awarded the project. The contractor's estimators or purchasing agents, both general and subcontractors, will work with many suppliers to obtain the most favorable pricing that meets the specifications' requirements. The manufacturer's representatives will use the specifications to ensure that products they produce for the general or subcontractors to use on the project meet those same requirements. The contractors are responsible for ensuring that all those providing materials or equipment for the project have the appropriate specifications for those products.

Inspectors for the governing municipality, city, county, or state, will use the specifications submitted to obtain the construction permit to ensure that the materials and workmanship specified for the project are being provided. If not, they will stop work on the project until all requirements are satisfied.

Fortunately, for each of the users of the specifications, there is a format that all of this construction information is assembled into, so they can find the information they are looking for quickly. Let's examine that structure.

THE SPECIFICATIONS STRUCTURE

The Construction Specifications Institute (CSI) established the basic format used in the specifications to help the construction industry consistently organize and present construction information. Originally, the CSI Master Format (1) broke all the written description of the project into 16 project divisions and 1 bidding division. Then, as buildings became more sophisticated, the need for more definition of low-voltage systems became necessary and the architectural community split them from the electrical division and created Division 17. This division contained all technology-related building systems and included all voice and data, access control, fire alarms and smoke detection systems, and other low-voltage systems. Subsequently, seeing that there was need for more succinct divisions, CSI created a new numbering system with 49 divisions. However, these divisions are more focused on larger projects such as dams, larger highway projects, manufacturing, and government projects, and the construction community found themselves divided into two groups—those who did large projects, approximately $100 million and larger, and those who did smaller projects, less than $100 million. The ones who did the larger projects adopted the 49 divisions, and those who did smaller projects adopted the 17 divisions.

The 17 divisions are shown in Figure 4.1 in a sample specification table of contents. Figure 4.2 is a sample of the 49 divisions table of contents.

Of the divisions, Division 1, "General Requirements," is the nontechnical division, while all other divisions are the technical divisions that describe the specific written requirements for the construction of the project. The important thing to remember about the divisions is that they always remain the same and always appear in the order shown. The division titles do not change, and they will always be in the same order in each project's set of specifications.

If you look at any specification book's table of contents, you will be able to verify that the divisions are the same and in the same order. Even if information is not needed from a division, the division's title will be listed in the table of contents with the note of "not used" or "not required." Notice that the titles of each division pertain to an element of construction and are in the basic sequence of construction for most projects. The structure of these divisions, always with the same titles, allows us to be able to pick up any project's specification book and immediately narrow our search to one division when seeking information, thus simplifying our search and saving valuable time, either in the bidding process or in the field.

If you look at any division in the specifications table of contents, you will see several topics listed beneath that are related to that division. For example, if I looked at Division 9, "Finishes," I would expect to find listed topics on carpet, tile, painting, and so on. Each one of these titles represents a section of that division. Each division will have multiple section titles listed in each division, depending on what is required for that project. If you look at Division 9 in the Master CSI Table of Contents at the end of this chapter you will see that Section 09900 pertains to painting. The first two digits represent the number of the division that section pertains to, and the last three digits are unique to that section. If a project does not have any painting, then that section will not be included in Division 9 of that project's specifications. Only the sections that are pertinent to the project will be included in the divisions for that project's specifications. Knowing this now allows us to narrow our information search to several pages within the specifications by determining in which section in the division the information we are seeking is contained. This numbering system also allows the architect to add and delete sections without affecting the placement and order of the sections.

17 DIVISIONS
TABLE OF CONTENTS

I. **PROPOSAL DOCUMENTS AND CONTRACT FORMS**
    AA - ADVERTISEMENT TO BIDDERS
    AB - SUB CONTRACTOR / SUPPLIER BID PROPOSAL FORM
    AD - PROPOSAL BOND
    AE - FELONY CONVICTION NOTIFICATION
    AG - AFFIDAVIT OF NON-DISCRIMINATORY EMPLOYMENT
    AH - AFFIDAVIT OF NON-ASBESTOS, LEAD, AND PCB USE IN PROJECT
    AI - WARRANTY FORM
    AJ - WAIVER AND RELEASE OF LIEN FORM
    AM - CONFLICT OF INTEREST QUESTIONNAIRE

    BA - CONTRACT DOCUMENTS

II. **SPECIFICATIONS**
DIVISION 1 - GENERAL REQUIREMENTS
SECTION   01010 - SUMMARY OF WORK
               01018 - OWNER PROVIDED DOCUMENTS
               01020 - ALLOWANCES
               01030 - ALTERNATES
               01050 - FIELD ENGINEERING
               01091 - CODES, REGULATIONS AND STANDARDS
               01110 - NOTIFICATION OF ARCHITECT REQUIREMENTS
               01115 - TESTING AND INSPECTION LABORATORY SERVICES
               01200 - PROJECT MEETINGS
               01300 - SUBMITTAL PROCEDURES
               01370 - SCHEDULE OF VALUES
               01390 - PAYMENT PROCEDURES
               01400 - QUALITY CONTROL
               01480 - CONSTRUCTION SCHEDULE
               01501 - TEMPORARY FACILITIES
               01631 - PRODUCTS AND SUBSTITUTIONS
               01710 - CLOSEOUT PROCEDURES
               01780 - PROJECT RECORD DOCUMENTS

DIVISION 2 - SITE WORK
SECTION   02010 - GEOTECHNICAL REPORT
               02112 - WASTE MATERIAL DISPOSAL
               02220 - TRENCHING AND BACKFILLING
               02230 - TRENCH SAFETY SYSTEM
               02240 - SOIL STABILIZATION
               02280 - TERMITE CONTROL
               02441 - IRRIGATION SYSTEM (BY ALLOWANCE)
               02480 - PLANTING (BY ALLOWANCE)
               02489 - HYDRO MULCH SEEDING (BY ALLOWANCE)
               02530 - POST-TENSIONED CONCRETE TENNIS COURTS
               02830 - PVC COATED CHAIN LINK FENCE AND GATES
               02874 - BICYCLE RACKS
               02900 - LANDSCAPE PLANTING
               02920 - TOPSOIL

FIGURE 4.1 Sample of Table of Contents for 17 Divisions

DIVISION 3 - CONCRETE
      03321 - LIGHTWEIGHT INSULATING CONCRETE DECK SYSTEM
      03490 - GLASS FIBER REINFORCED CONCRETE
      03530 - CEMENTITIOUS WOOD FIBER ROOF SYSTEM
      03540 - SELF-LEVELING UNDERLAYMENT CONCRETE

DIVISION 4 – MASONRY
      04200 - UNIT MASONRY
      04720 - ARCHITECTURAL CAST STONE

DIVISION 5 – METALS
      05410 - COLD FORMED METAL FRAMING
      05500 - MISCELLANEOUS METALS
      05515 - ALTERNATING TREAD STAIR
      05800 - EXPANSION CONTROL

DIVISION 6 - WOOD AND PLASTICS
SECTION   06100 - ROUGH CARPENTRY
      06220 - FINISH CARPENTRY AND MILLWORK

DIVISION 7 - THERMAL AND MOISTURE PROTECTION
SECTION   07115 - SHOWER STALL WATERPROOFING
      07140 - BELOW GRADE WATERPROOFING
      07160 - DAMPPROOFING ABOVE GRADE
      07210 - BUILDING INSULATION
      07250 - SPRAYED-ON FIREPROOFING
      07270 - FIRESTOPPING AND FIRE SAFING
      07402 - PREFINISHED METAL ROOFING
      07525 - MODIFIED BITUMEN MEMBRANE ROOFING SYSTEM (ALTERNATE)
      07530 - COAL-TAR ELASTOMERIC MEMBRANE (CTEM) BUILT-UP ROOFING SYSTEM (BASE BID)
      07620 - ROOF RELATED SHEET METAL
      07650 - FLEXIBLE FLASHING
      07721 - ROOF ACCESSORIES
      07831 - ROOF SCUTTLE
      07900 - BUILDING SEALANTS

DIVISION 8 - DOORS AND WINDOWS
SECTION   08100 - HOLLOW METAL DOORS AND FRAMES
      08213 - PLASTIC LAMINATE FACED WOOD DOORS
      08301 - MUSIC ROOM DOOR AND FRAME ASSEMBLIES
      08332 - OVERHEAD ROLLING GRILLES
      08710 - DOOR HARDWARE
      08750 - DOOR THRESHOLDS AND SEALS
      08800 - GLAZED SYSTEMS
      08950 - INSULATED TRANSLUCENT SKYLIGHTS

DIVISION 9 - FINISHES
SECTION   09100 - LATH AND PLASTER
      09250 - GYPSUM WALLBOARD SYSTEMS
      09310 - CERAMIC TILE
      09311 - PORCELAIN TILE
      09330 - QUARRY TILE
      09510 - ACOUSTICAL LAY-IN CEILING
      09520 - ACOUSTICAL WALL PANELS, FABRICS AND DIFFUSERS.
      09550 - WOOD GYMNASIUM FLOORING

Table of Contents - 2

FIGURE 4.1 (Continued)

FIGURE 4.1 (Continued)

FIGURE 4.1 (Continued)

FIGURE 4.1 *(Continued)*

49 Divisions
TABLE OF CONTENTS

I. **PROPOSAL DOCUMENTS AND CONTRACT FORMS**

AA - BID FORM
AB - SUMMARY INFORMATION FOR SUBCONTRACTORS AND MATERIAL SUPPLIERS
AC - COMPETITIVE SEALED PROPOSAL FORM
AD - PROPOSAL BOND
AE - FELONY CONVICTION NOTIFICATION
AF - AFFIDAVIT OF NON-DISCRIMINATORY EMPLOYMENT
AG - LIST OF SUBCONTRACTORS
AH - AFFIDAVIT OF NON-ASBESTOS, LEAD, AND PCB USE IN PROJECT

BA - CONTRACT DOCUMENTS
BB - TEXAS STATUTORY PERFORMANCE BOND
BC - TEXAS STATUTORY PAYMENT BOND

II. **SPECIFICATIONS**

DIVISION 1 - GENERAL REQUIREMENTS

SECTION
01 11 00 - SUMMARY OF WORK
01 21 00 - ALLOWANCES
01 22 00 - MEASUREMENT AND PAYMENT (UNIT PRICES)
01 23 00 - ALTERNATES
01 25 13 - PRODUCT SUBSTITUTION PROCEDURES
01 26 00 - CONTRACT MODIFICATION PROCEDURES
01 29 00 - PAYMENT PROCEDURES
01 29 73 - SCHEDULE OF VALUES
01 31 13 - PROJECT COORDINATION
01 31 19 - PROJECT MEETINGS
01 32 16 - CONSTRUCTION PROGRESS SCHEDULE
01 33 00 - SUBMITTAL PROCEDURES
01 35 16 - ALTERATION PROJECT PROCEDURES
01 41 00 - REGULATORY REQUIREMENTS
01 45 00 - QUALITY CONTROL
01 45 23 - TESTING AND INSPECTING LABORATORY SERVICES
01 50 00 - TEMPORARY FACILITIES AND CONTROLS
01 71 23 - FIELD ENGINEERING
01 73 29 - CUTTING AND PATCHING
01 74 19 - CONSTRUCTION WASTE MANAGEMENT AND DISPOSAL
01 77 00 - CLOSEOUT PROCEDURES
01 78 39 - PROJECT RECORD DOCUMENTS
01 91 00 - COMMISSIONING

DIVISION 2 - EXISTING CONDITIONS

SECTION
02 32 00 - GEOTECHNICAL REPORT
02 41 13 - SELECTIVE DEMOLITION

Table of Contents - 1

FIGURE 4.2 Sample of Table of Contents, 49 Divisions

Master Format® 2004 Division Numbers and Titles and Master Format™ 2010 Titles Group, Subgroups, and Divisions used in this book are published by The Construction Specifications Institute (CSI) and Construction Specifications Canada (CSC); used with permission from CSI. For those interested in a more in-depth explanation of Master Format 2010 and its use in the construction industry, visit www.csinet.org/masterformat or contact: The Construction Specifications Institute, 110 South Union Street, Suite 100, Alexandria, VA 22314, www.csinet.org.

<u>DIVISION 3 - CONCRETE</u>

SECTION   03 30 00 - CAST-IN-PLACE CONCRETE
                        03 32 00 - LIGHTWEIGHT INSULATING CONCRETE DECK SYSTEM
                        03 35 19 - STAINED CONCRETE

<u>DIVISION 4 - MASONRY</u>

SECTION   04 20 00 - UNIT MASONRY
                        04 72 00 - ARCHITECTURAL CAST STONE

<u>DIVISION 5 - METALS</u>

SECTION   05 12 00 - STRUCTURAL STEEL
                        05 21 00 - OPEN WEB STEEL JOISTS AND JOIST GIRDERS
                        05 31 00 - STEEL DECK
                        05 40 00 - COLD FORMED METAL FRAMING
                        05 50 00 - METAL FABRICATIONS
                        05 51 13 - PRE-ENGINEERED STEEL PAN STAIRS
                        05 51 33.23 - ALTERNATING TREAD STAIR

<u>DIVISION 6 - WOOD AND PLASTICS</u>

SECTION   06 10 00 - ROUGH CARPENTRY
                        06 20 00 - FINISH CARPENTRY AND MILLWORK
                        06 61 16 - SOLID POLYMER FABRICATIONS

<u>DIVISION 7 - THERMAL AND MOISTURE PROTECTION</u>

SECTION   07 11 00 - DAMPPROOFING ABOVE GRADE
                        07 13 00 - SHOWER STALL WATERPROOFING
                        07 14 00 - WATERPROOF COATING
                        07 16 00 - BELOW GRADE WATERPROOFING
                        07 21 00 - THERMAL INSULATION
                        07 24 00 - EXTERIOR INSULATION AND FINISH SYSTEMS
                        07 42 13 - METAL WALL PANELS
                        07 42 16 - PRE-FINISHED ALUMINUM FABRICATIONS AND COMPOSITE PANELS
                        07 52 50 - MODIFIED BITUMEN "COOL ROOF" MEMBRANE ROOFING SYSTEM
                        07 62 00 - ROOF RELATED SHEET METAL
                        07 72 10 - ROOF ACCESSORIES
                        07 72 33 - ROOF SCUTTLES (HATCHES)
                        07 81 00 - SPRAYED-ON FIREPROOFING
                        07 84 00 - FIRESTOPPING AND FIRE SAFING
                        07 92 00 - JOINT SEALANTS
                        07 95 00 - EXPANSION CONTROL

<u>DIVISION 8 - DOORS AND WINDOWS</u>

SECTION   08 11 13 - HOLLOW METAL DOORS AND FRAMES
                        08 14 23.16 - PLASTIC LAMINATE FACED WOOD DOORS
                        08 31 13 – ACCESS DOORS
                        08 33 13 - COILING COUNTER DOORS
                        08 33 23 - OVERHEAD COILING DOORS
                        08 42 29.13 – MONUMENTAL ALUMINUM FRAMED FOLDING PANEL SYSTEM
                        08 80 00 - GLAZED SYSTEMS

Table of Contents - 2

FIGURE 4.2 *(Continued)*

08 91 00 - ALUMINUM LOUVERS

DIVISION 9 - FINISHES

SECTION   09 21 16 - GYPSUM WALLBOARD SYSTEMS
                 09 24 00 - LATH AND PLASTER
                 09 30 13 - CERAMIC TILING
                 09 30 19.13 - PORCELAIN TILE
                 09 51 00 - ACOUSTICAL LAY-IN CEILING
                 09 54 23.13 - LINEAR METAL SOFFIT PANELS
                 09 61 13 - SLIP RESISTANT FLOORING TREATMENT
                 09 65 19 - RESILIENT TILE FLOORING AND BASE
                 09 67 23 - RESINOUS FLOORING
                 09 68 00 - CARPET
                 09 91 00 - PAINTING AND STAINING
                 09 97 23 - CONCRETE FLOOR SEALER

DIVISION 10 - SPECIALTIES

SECTION   10 01 00 - MISCELLANEOUS SPECIALTIES
                 10 11 00 - MARKERBOARD AND TACKBOARD
                 10 12 00 - DISPLAY CASES
                 10 14 00 - GRAPHICS
                 10 21 13.19 - SOLID POLYMER TOILET PARTITIONS, SHOWER COMPARTMENTS AND BENCHES
                 10 22 26.13 - ACOUSTICAL ACCORDION FOLDING PARTITIONS
                 10 28 13 - TOILET ACCESSORIES
                 10 44 13 - FIRE EXTINGUISHER CABINETS
                 10 51 53 - METAL ATHLETIC LOCKERS AND BENCHES
                 10 73 16 - ALUMINUM CANOPY
                 10 75 00 - ALUMINUM FLAGPOLE
                 10 81 13 - BIRD CONTROL

DIVISION 11 - EQUIPMENT

SECTION   11 31 00 - RESIDENTIAL APPLIANCES
                 11 40 00 - FOOD SERVICE EQUIPMENT
                 11 68 33 - OUTDOOR ATHLETIC EQUIPMENT

DIVISION 12 - FURNISHINGS

SECTION   12 21 13 - HORIZONTAL BLINDS
                 12 24 00 - WINDOW SHADES
                 12 35 50 - EDUCATIONAL CASEWORK
                 12 93 13 - BICYCLE RACK

DIVISION 13 - SPECIAL CONSTRUCTION

                 NOT USED

DIVISION 14 - CONVEYING SYSTEMS

SECTION   14 24 00 - HYDRAULIC ELEVATORS
                 14 24 10 - MACHINE ROOM-LESS ELEVATORS

Table of Contents - 3

FIGURE 4.2 *(Continued)*

DIVISION 21 - FIRE SUPPRESSION

SECTION   21 11 00 - WET-PIPE FIRE SUPPRESSION SPRINKLERS

DIVISION 22 - PLUMBING

SECTION   22 05 17 - SLEEVES AND SLEEVE SEALS FOR PLUMBING PIPING
           22 05 29 - PLUMBING HANGERS AND SUPPORTS
           22 11 16 - DOMESTIC WATER PIPING
           22 13 16 - SANITARY WASTE AND VENT PIPING
           22 14 13 - STORM DRAINAGE PIPING
           22 30 00 - PLUMBING EQUIPMENT
           22 40 00 - PLUMBING FIXTURES

DIVISION 23 - MECHANICAL

SECTION   23 05 13 - COMMON MOTOR REQUIREMENTS FOR HVAC EQUIPMENT
           23 05 17 - SLEEVES AND SLEEVE SEALS FOR HVAC PIPING
           23 05 18 - ESCUTCHEONS FOR HVAC PIPING
           23 05 19 - METERS AND GAUGES FOR HVAC PIPING
           23 05 23 - GENERAL DUTY VAVLES FOR HVAC PIPING
           23 05 29 - HANGERS AND SUPPORTS FOR HVAC PIPING AND EQUIPMENT
           23 05 33 - HEAT TRACING FOR HVAC PIPING
           23 05 48 - VIBRATION CONTROLS FOR HVAC PIPING AND EQUIPMENT
           23 05 53 - IDENTIFICATION FOR HVAC PIPING AND EQUIPMENT
           23 05 93 - TESTING, ADJUSTING, AND BALANCING FOR HVAC
           23 07 13 - DUCT INSULATION
           23 07 19 - HVAC PIPING INSULATION
           23 09 00 - INSTRUMENTATION AND CONTROLS FOR HVAC
           23 11 23 - GAS PIPING
           23 21 13 - HYDRONIC PIPING
           23 21 23 - HYDRONIC PUMPS
           23 23 00 - REFRIGERANT PIPING
           23 25 00 - HVAC WATER TREATMENT
           23 31 13 - METAL DUCTS
           23 33 00 - AIR DUCT ACCESSORIES
           23 34 00 - HVAC FANS
           23 36 00 - AIR TERMINAL UNITS
           23 41 00 - PARTICULATE AIR FILTRATION
           23 64 26 - ROTARY-SCREW WATER CHILLERS
           23 73 13 - MODULAR INDOOR CENTRAL-STATION AIR HANDLING UNITS
           23 74 15 - PACKAGED COMMERCIAL ROOFTOP UNITS LESS THAN 25 TONS
           23 81 26 - SPLIT-SYSTEM AIR-CONDITIONERS
           23 82 39 - UNIT HEATERS
           23 83 23 - RADIANT-HEATING ELECTRIC PANELS

DIVISION 25 – INTEGRATED AUTOMATION-

           NOT USED

DIVISION 26 - ELECTRICAL

SECTION   26 05 00 - BASIC ELECTRICAL MATERIALS AND METHODS
           26 05 01 - ELECTRICAL UTILITY SERVICES - UNDERGROUND
           26 05 19 - LOW-VOLTAGE ELECTRICAL POWER CONDUCTORS AND CABLES

FIGURE 4.2 (Continued)

FIGURE 4.2 *(Continued)*

END OF LIST

Table of Contents - 6

FIGURE 4.2 ***(Continued)***

Once we have identified the division and section we need to look at for information, we can further expedite our search if we understand that each section has three parts: general, product, and execution. This is where we define why we are looking in the specifications in the first place. The types of information presented to us in the specifications are quality, materials, and methods. *Quality* pertains to the workmanship, references, scheduling, protection of work-in-progress, and contractor qualifications and will be found in Part One, *General*, of each section. *Materials* pertains to the specific materials allowed for use to construct the project, their quality, grade, finish, fabrication, and acceptable manufacturers, and can be found in Part Two, *Product*, of each section. *Methods* to be utilized by the contractor for the functions of work, such as inspection, preparation for placement, and adjusting for each section are found in Part Three, *Execution*.

Now that we know the structure used for the specifications, we can find information quickly without having to spend time searching page after page. Because we know that the specifications have the same divisions and that each division has various sections, we can go to the table of contents for a division and find the section with the title that best matches the information we are seeking. Then, because we know that the information contained in the specifications is quality, materials, and methods, we can go to the correct part of the section to see the information we are looking for. We will spend more time with this process in Chapter 5.

INFORMATION IN THE SPECIFICATIONS

Specifications contain many important pieces of construction information that pertain to the project. By looking at the divisions, we can see how this information is grouped.

Division 0

This division is included at the beginning of the specifications and typically includes all the information needed to submit a bid or proposal for completing the project and any other information for administering the project, such as the general conditions, which do not specifically pertain to the quality, material, and methods required.

THE BIDDING DOCUMENTS

The bid documents include all the instructions for submitting bids for a particular project. This bidding information includes the invitation to bid, instructions to the bidders, and the bid forms. These documents will only contain information for the project whose specification book they are included in. This information and the forms will be different for each construction project.

The invitation to bid contains information about the project, the owner and architect, who is invited to bid, and when and where the bids must be submitted.

The instructions to the bidders indicate all the information needed to correctly complete the bid forms and properly submit them. These instructions will define the bonding requirements, contractor qualifications and disclosures, and how to complete the forms. The instructions must be followed explicitly in order for the bid to be considered. The bidding requirements and submittal date and time must be met exactly as indicated. Any deviation will result in having your bid disqualified.

The bid forms must be completed exactly as indicated, including certification that all addenda have been considered in the amount being submitted to complete the project. The forms shown in the specifications must be used to submit the bid. No other forms will be accepted, and these forms will only pertain to that specific project.

THE GENERAL CONDITIONS OF THE CONTRACT FOR CONSTRUCTION
The *General Conditions of the Contract for Construction* are purchased off the shelf and are typically used in their entirety for the construction project with little or no modifications, as discussed in Chapter 2. However, only occasionally will they be reprinted in the specifications. When they are, they will typically be found in Division 0. When not included in their entirety, there will be short section in Division 0 that states they are included by reference. We will see what that means as we look at the definitions of the technical specifications.

DIVISION 1 GENERAL REQUIREMENTS
This division's sections contain what are considered the nontechnical specifications of the project and outline the general requirements that will be used to govern the other divisions. This information will vary from project to project but will typically include the scope of work and information such as requirements for the site project signs, project clean-up, start-up of new equipment, clean-up of the site, and warranties. Also included will be the supplementary conditions. It is usual, because of each construction project's unique requirements, to have a need for supplementary conditions to make significant modifications or additions to the general conditions. These supplementary conditions will be written referencing the AIA document and will create conditions and requirements that are specific only to that particular project.

Divisions 2 through 49

These divisions' sections are considered the technical specifications and delineate the specific requirements for the materials that are allowed on the project, the quality controls for the construction process, and the methods allowed in the functions of completing the work. They establish the type and quality of materials, equipment, and fixtures used on the project, the quality of workmanship required for the project, and the methods of fabrication, installation, and erection. Also included will be references to many documents that will be used to complete the project but that are not printed in the specifications. These are called *reference documents*.

It is not uncommon for specified requirements to be derived from another organization's written standards. The specifications will note that some aspect of the work must be executed as specified in the document created by the originating organization. However, instead of including that entire document or the appropriate section applicable to this project, it is included in the specifications by reference. If work is specified to be completed as defined in a reference document, the contractor must complete the work as though the definition were printed in the specifications for this project. If the contractor is not sure what the referenced document says about that aspect of the work, he must obtain a copy of that document and complete the work as indicated.

The new-format specifications are divided into subgroups. These allow us to classify the work more closely with a buildings function. The subgroups are construction, services, site and infrastructure, and process equipment. Those familiar with the old 16 divisions will note that they exist in the facility construction subgroup and are intact in the new format with a few changes. Division 2, previously called "Sitework," is now labeled "Existing Conditions" and sitework is covered in the "Site and Infrastructure" subgroup. Division 8, previously "Doors, Windows and Glass," is now simply labeled "Openings." The most significant changes come in Divisions 15 and 16, formerly "Mechanical" and "Electrical." Those sections listed for those two divisions are broken out into distinct groups of fire protection; plumbing; heating, ventilating, and air conditioning; integrated automation; electrical; communications; and electronic safety and security in the "Facility Services" subgroup. The "Process Equipment" subgroup deals with the installation of building processes and the related equipment. The new format makes it much easier to identify elements of a particular system and save time, whether in the office estimating or constructing in the field.

Division 17 (Old Format)

This division is used in the old 16-division format,* and used when there is a significant number of low-voltage systems in a project that would complicate the electrical construction for the project. Because these systems have special installation requirements, it is becoming a more common practice to separate these systems from the electrical to provide clearer definition. These systems usually include security systems, fire protection/detection systems, data and communications systems, and any other control-type systems. These are contained in their own divisions in the new format specifications, 49 divisions.

---

* There were originally 16 divisions recognized by CSI, but the industry created a 17th out of necessity; CSI never recognized the addition.

SUMMARY

We have looked at the structure of the specifications and have learned that there are multiple divisions and that they are always the same divisions in the same order. We have also learned that there can be multiple sections in each division, depending on the requirements of the project. Each section is divided into three parts—General, Product, and Execution—which coincide with the type of information found in the specifications, quality, materials, and methods. We now need to study how we can use the structure of the specifications and working drawings to quickly find construction information.

■ CHAPTER 4 LEARNING OUTCOMES

- Construction information is cataloged into divisions in each set of project specifications. These divisions are always titled the same, and always in the same order.

- The information provided in the specifications is related to quality, materials, and methods that pertain to the construction of a project.

- Each division has information on different topics listed by titles, which are called sections.

- The section has three parts—General, Product, and Execution, which correspond to the quality, materials, and methods information contained in the specifications.

- The sections in each division vary from project to project, depending on the projects construction requirements.

■ CHAPTER 4 QUESTIONS

1. Since the divisions in a specification book are always the same, list the 17 divisions, in the correct order.

2. The specifications provide what elements for interpreting the information contained in the divisions?

3. How are the divisions divided into different pieces of information?

4. The sections are divided into how many parts, and what are the names of the parts?

5

FINDING CONSTRUCTION INFORMATION IN THE WORKING DRAWINGS AND SPECIFICATIONS

OVERVIEW

Now that we understand where the working drawings and specifications evolved from and how they are structured, we are ready to begin practicing the methods used in determining where and how to look for specific construction information. We will study a method that consists of asking ourselves some questions about key words that describe the information we are seeking, in order to minimize the time needed to complete our search. We will also look at how the two documents, the working drawings and specifications, complement each other and are used together. We will study a process for use when looking at the documents for the first time. Before we proceed, though, let's review the documents structure.

DOCUMENT STRUCTURE REVIEW

In this review, we want to refamiliarize ourselves with the structure of the working drawings and specifications, because understanding this structure

Print and Specifications Reading for Construction, Updated Edition. Ron Russell.
© 2024 John Wiley & Sons, Inc. Published 2024 by John Wiley & Sons, Inc.
Companion website: www.wiley.com/go/printspecreadingupdatededition

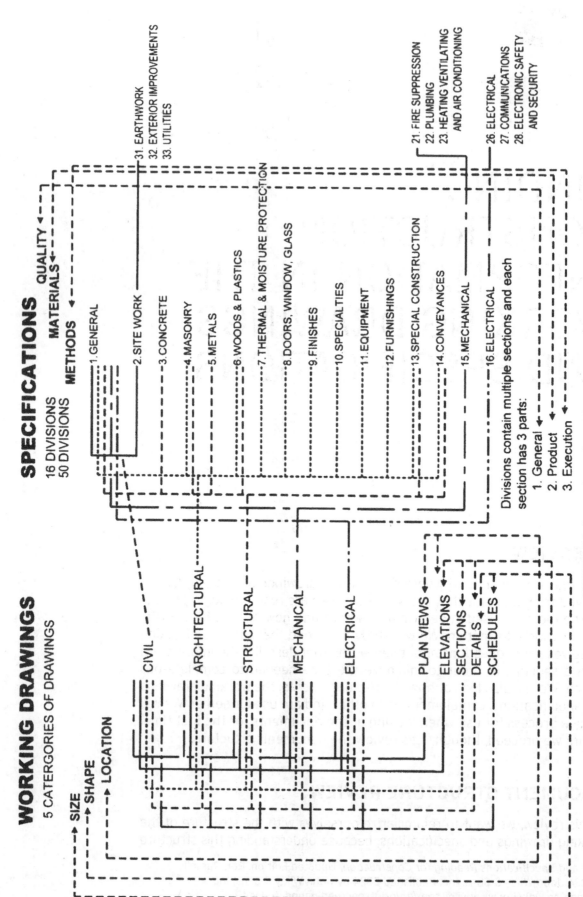

FIGURE 5.1 Comparison of Drawings and Specs

is key to reducing the amount of time required to locate information in these documents. Figure 5.1 provides an illustration of what the structures look like.

Working Drawings

As discussed in Chapter 3, the working drawings comprise many sheets grouped into six categories: civil, architectural, structural, mechanical, electrical, and plumbing. The number of sheets in a drawing set will vary, depending on the complexity of the project and the amount of detail provided by the architect and engineers. In each of the categories, we will find examples of the five types of drawings: plan views, elevations, sections, details, and schedules.

The most important thing we need to know about the drawings is that the type of information shown in the drawing package is information relating to the size, shape, and location of items used to construct the project. Most of the size information and dimensions will be found on detail and schedule drawings, as well as on the others; the shape information can be found on elevations, sections, and detail drawings, and the location information can be found on the plan views and elevations.

Specifications

The specifications are broken down into divisions as established by the Construction Specifications Institute. There are multiple divisions, and they are always the same divisions and are always in the same order. They do not change. In the 49 division specifications, Division 0 contains the bidding information, Division 1 is considered the nontechnical division, and Divisions 2 through 49 are considered the technical divisions.

In each division, there are units of information on various construction topics called sections. The sections in each division change from project to project, depending on the scope of work and building design. Each section has three parts, General, Product, and Execution.

It is important to know that the specifications provide information on the quality, materials and methods used to construct the project. This corresponds with the three parts of the section, quality information being pre-sented in the *General* part of the section, materials information being presented in the *Product* part of the section, and methods information being presented in the *Execution* part of the section.

It is also important to note, as shown in Figure 5.1, that certain categories of the drawings correspond the divisions of the specifications. Division 1 pertains to all the drawing categories and each of the subsequent divisions in the specifications. However, if I am seeking information regarding the civil work or sitework, I would probably have the specification book open to Division 31 through 35 and the drawings opened to the civil drawings. If I were looking for information related to the HVAC, plumbing, supply piping, or fire protection systems, I would have the specifications open to Divisions 21 through 28 and the drawings turned to the mechanical or plumbing category of the drawings. By understanding this relationship between these two documents, you will expedite the search for information.

FIRST LOOK AT THE WORKING DRAWINGS AND SPECIFICATIONS

When you first receive a new set of working drawings and specifications for a project, you will need to spend some time with them to become familiar with the project requirements. This could be an hour to a 3-hour process, depending on the project. The intent is to gain a basic understanding of the project and the construction materials and methods that would be employed to build it. By following a prescribed method for first examining the documents, you will gain a general understanding of the size of the project, basic building components used, and some of the construction techniques to be employed. This first look will also facilitate locating information in these documents quickly. Let's look at the method for examining the working drawings (see Table 5.1).

The first sheet in the working drawing package to examine will be the cover sheet, or first sheet. Sometimes, this sheet has a rendering or drawing of the building—typically, a copy of the presentation drawing used to show the project to the interested parties, which is an excellent tool for getting a perspective on what the finished building would look like. If there is no picture of the building, then this drawing should be reviewed to identify the name of the project, who the architect and owner are, who the significant contractors are, and other descriptive information regarding who is involved and where the project is located.

The next sheet to examine is called the index of drawings. This sheet will list all the sheets included in each of the drawing categories. You should also look to see if additional categories were used—that is, was the landscaping broken out of the civil category and given a category of its own because there was a significant amount of information required for this part of the project? The index of drawings will show all of the drawing categories and will list each sheet included in those categories. At this point, you should go through the entire drawing package checking sheet numbers to verify that you have all of the individual sheets listed on the index. If you have sheets missing, now is the time to contact the architect

TABLE 5.1 Reading the Working Drawings

READING WORKING DRAWINGS
REVIEWING THE STEPS IN READING A SET OF WORKING DRAWINGS
STEP 1 - STUDY TITLE PAGE AND INDEX OF DRAWINGS
STEP 2 - START WITH THE ARCHITECTURAL DRAWINGS
STEP 3 - STUDY PLAN VIEW AND INTERIOR ELEVATIONS
STEP 4 - STUDY SECTIONS
STEP 5 - STUDY DETAILS
STEP 6 - STUDY EXTERIOR ELEVATIONS
STEP 7 - NOTE THE DIFFERENT TYPE OF SCHEDULES
STEP 8 - READ GENERAL NOTES
STEP 9 - DETERMINE THE TYPE OF STRUCTURAL SYSTEM AND COMPONENTS
STEP 10 - STUDY EACH OF THE MEP CATEGORIES ONE AT A TIME
STEP 11 - STUDY THE CIVIL AND SITE PLANS
STEP 12 - STUDY THE MEP SITE DRAWINGS

and obtain copies. You should not begin using this document until you are sure you have all of the sheets. Once you have completed looking at the index of drawings, you can proceed to examine the sheets in the different categories, starting with the architectural category.

Examination of the sheets in the architectural category will provide the broadest understanding of the building definition, use of materials, and construction techniques. In examining any of the categories, it is best to begin with the plan views. This will allow you to see how the building interior is arranged and how intended people and material flows. You can also look at the callout symbols used to allow you to move from the plan views to the elevations, section, and details. Study the plan views first and then begin using the callouts to examine the elevations, sections, and details, in that order. Looking at the elevations will give you a perspective on the outside appearance and exterior materials used. Then, progressing through the sections and details, you can move deeper into the building design and understand construction methods.

Once comfortable with the information on these sheets, you should look at the schedules in the architectural category and identify what symbols are used to denote which items on the types of drawings—that is, circles for doors and squares for windows. Reviewing the schedules will also give you some ideal of quantities of these items that will be required to complete the project. The last thing to examine would be the notes on each sheet to complete the general understanding of this category of the drawings. The overall review of this category should leave you with an understanding of the building configuration, types of finishes used, type of roofing system, exterior materials, and overall size of the building.

Once you have completed a review of the architectural category and understand the basics of the building use and construction, you should progress to the structural, mechanical, and then electrical categories of the drawing package. The structural category should reveal the types of structural systems used—either steel, concrete, masonry, wood, or some combination of those types. The review of the mechanical and plumbing categories should give you some definition of the type of supply process piping systems, drain systems, fire protection systems, and HVAC systems. Review of the electrical category should provide information on the electrical distribution system throughout the building and allow you to understand switchgear placement and panel locations. In each category review, you should follow the same basic approach to examining the information, starting with the plan views, and then moving to the elevations, sections, details, and finally with the examination of any schedules and notes.

After reviewing these categories, to complete your understanding of the building you must now examine the civil drawings to understand how the building is oriented to the property. Most of the drawing sheets in this category will be plan views. A review of these plan views will indicate the locations of utilities, paving and parking areas, and how the site must be graded for drainage. Plan views will coordinate with the other categories. After completing this review, you should have a good understanding of the overall project's size, shape, and location of building components and materials. Once you are familiar with the working drawings, you can examine the specifications to obtain a clearer picture of the quality, materials, and methods to be used to construct the building.

As with the working drawings, when we first get the specifications we need to examine them for completeness and to get a basic understanding of what is included in the project. It is not necessary to read every page and paragraph at this time.

The first thing to verify is that on the cover page there is a reference to the same project as on the working drawings. You must be sure that the set of specifications you are working with are for the same project as the working drawings. Verify this by confirming the project title and architects' names are on the cover page.

Next, we should look at the table of contents first making sure that there is a listing of all the divisions. Note which divisions have sections and which are marked "not used." This will tell you what elements of construction are included in the project and which are not. Next, starting with Division 2, look at the titles of each section listed for each division. Try to remember key words to these titles so that as you work through the documents you will recall having seen that there is information available on topics for which you are seeking construction information. Next, thumb through the specifications slowly while looking at the section numbers printed at the bottom of the pages. Make sure that you see at least one page from each section listed for the divisions in the table of contents. If you do not see at least one page for a section, stop and confirm whether you have that section and contact the architect to obtain a copy of the section if it is missing.

Finally, examine the sections in Division 1. Each of these sections should be scanned thoroughly to gain a good understanding of what is required for completing the other sections.

Once you have completed this review of the working drawings and specifications, you are ready to start using them to find the information you need to construct the project. Now let's look at a method for searching for this information.

FINDING CONSTRUCTION INFORMATION IN THE WORKING DRAWINGS AND SPECIFICATIONS

When searching for construction information in the working drawings and specifications, you have to become accustomed to developing and identifying *key words*. Key words will enable you to take maximum advantage of your understanding of the documents structures to locate the information you are seeking.

To begin your search, the first step is to decide which document you should look at for the information. The object of being correct with this decision is to eliminate wasted time looking in the specifications when the information you seek is in the working drawings. The method that will allow you to minimize the likelihood of wasting this time, is to determine what is the key word that describes what you are looking for in terms of size, shape, and location, and quality, materials, or methods. If you decide the information you are looking for is related to size, shape, or location, you can put down the specifications and begin searching in the working drawings since that is the document that provides us information on size, shape, and location.

If the key word you are using is related to quality, materials, or methods, then you can set aside the working drawings and start your search in the specifications to answer this, since that is the document that provides us with that type of information. For example, if you are looking for information regarding "where the electrical room is located in the building," the key word in this statement is *located*. Since the drawings show location, you would put the specifications aside and begin your search with the working drawings. If you are looking for information on "what species of wood is required for the chair railings," you could restate this to say "what material is required for the chair railings." Obviously, *species of wood* are the key words, and this is a *materials* question. You would put aside the working drawings and begin your search in the specifications, since this is the document that provides you with material specifications. If you can successfully make the determination of which document to begin your search with by using key words, you will have automatically reduced your search time by 50 percent by eliminating half of the information you must review to find what you are looking for. Once you have determined which document you are going to use, the structure of that document can be utilized to quickly reduce your search time even more.

For example, let's take the question of "where the electrical room is located in the building." We determined that this information would be found in the working drawings, since that is the document that provides location information. We set the specifications aside and pick up the plans. We could start at the very beginning of the working drawings, but that would waste a lot of time before we came upon the information we are seeking. Since we know the structure of the drawings, we can better our search time by again using key words. We are looking for the electrical room. We know from having studied them that the working drawings have a category of sheets called the electrical category. We also learned in our studies that room locations are shown on plan views. As a result, the first logical step would be to go to the index of drawings and look at all the plan view title sheets listed in the electrical category and find one with the key words *electrical plan view*. Now, bear in mind that it is possible, even probable, that the electrical room will be also be shown on a plan view in the architectural category. However, there is much more information on that plan view that could substantially inhibit your search for the information and cost valuable time. It would be much more efficient to look on the plan views in the electrical category, where you can turn directly to a plan view that contains only electrical information, and quickly identify the electrical room, since most of your conduit runs would obviously originate from that single location on the drawing.

Using this thinking process, or key word method, will reduce your search time dramatically versus thumbing through each sheet to locate the information.

Take the example of "finding what species of wood is required for the chair railings." We have determined that the key word in this statement is *species of wood*, which corresponds with materials, which is the type of information provided for us in the specifications. Thus, you would set aside the working drawings and pick up the specifications to begin your search. The first thing you should recall about this document is the structure and that there are 16 divisions, one of these being Division 6, "Woods and Plastics." One of the key words is *species of WOOD*. Therefore, you would turn to Division 6, thus reducing the amount of information you must look through by 94 percent. You would turn to the table of contents, find the Division 6 listing of section titles, and select the title that most closely fits the key words you are using. For example, you would look for a section title worded something like *wood trim* or *finish wood trim*. Since you know that chair railings are considered a finish wood trim piece, either one of these titles would indicate that you could find the information you are seeking in that section. Once you select the section you are going to look in, and because you have learned that each section has three parts– General, Product, and Execution–you can determine in which part of the section to begin your search with. Part 2 of a section, Products, tells us about the materials used on the project, so you should be able to turn to Part 2 of that section to begin looking for the information on the species of wood required for the chair railings.

By using this key word process and with the knowledge of the structures of the documents, you can dramatically reduce the amount of time required to locate information in the working drawings and specifications. We must also remember that this process could require the use of both documents for the same key words.

For example, what if we needed to know what hardware to order for all doors that lead from the office area to the shop floor of a project? We would need both documents. The architect has denoted doors on the plans using letters in circles, and doors are used in multiple locations on the project. The first thing we need to determine is which of the doors allows access from the offices to the shop floor. In order to determine this, we must decide which document to look at first.

The key word for the information we are seeking is *location*. We must first see which doors are "located" in walls between the offices and the shop floor. To do this, we must examine the drawings since they are the documents that provide us with location information. In the drawings, we know there is a category called *architectural* that will show us wall, door, and window locations. We also know that these locations will probably be shown on a plan view, so we should turn to the index of drawings and look for a sheet title in the architectural category that has some wording that matches what we are looking for: *office floor plan, shop plan view,* or *first floor plan view*. Once we identify the correct plan view, we would identify the callout symbol for doors that allow access between the offices and the shop floor. Once we know the callout symbol for these doors, we are ready for the next step in determining what hardware is needed.

Usually, the plan views will not list all of the hardware for each door. But, we know that the doors are listed on a schedule type drawing, hence the use of the callout symbol, and that the schedule drawing sheet will tell us most of the information we will need for each of the doors. This method is used so the information will not clutter the plan view. We should now identify the schedule sheet drawing number in the index of drawings and turn to that sheet. The door schedule will list things like the door jamb type, whether the door is metal or wood, what type of finish is required, and what hardware package is to be used with the door. The hardware package will probably only be identified by a number or letter, depending on what the architect selected as the identifier. On the schedule, we would be able to see that doors D use hardware package "4." Now we are looking for something different; our key word is now materials, which is what the door hardware is. To understand what is called for in hardware package "4," we must look to the specifications.

Again, using the structure of the specifications, we know that of the 16 divisions, Division 8 is always titled "Doors, Windows, and Glass." Therefore, we can immediately turn to the specifications' table of contents and look at all the sections listed under Division 8 for a section with a title that most likely suggests it would contain information on door hardware. Upon turning to that section, we would look at Part 2, Product, to discover all the parts included in Hardware Package 4—that is, lockset, strike, closure, hinges, and so on.

To answer our original question, we had to begin our search with key words that led us through the working drawings using plan views and schedules in the architectural category. After obtaining the first part of information needed, our new keyword led us to a section in Division 8 in the specifications. By utilizing this method, significant time can be eliminated from your information search, even if you have never before been exposed to what you are looking for.

SUMMARY

The structure of the documents and the decision of which document to use first—the plans for size, shape, and location, or the specifications for quality, materials, and methods will dramatically reduce your search times for construction information in the working drawings and specifications. Identifying key words correctly will allow you to narrow your information search to specific areas of the documents. Knowing the structures of the documents will allow you to go quickly to the areas of the documents that contain information you are seeking. Now we will practice this method of finding information to become proficient in its use.

■ **CHAPTER 5 LEARNING OUTCOMES**

- Divisions in the specifications correspond to categories in the working drawings.

- Developing key words is important for determining where to look in the drawings and specifications for information.

■ **CHAPTER 5 QUESTION**

1. What divisions correspond what with what categories of drawings?

SECTION

BUILDING THE PROJECT

This section will provide a look at the materials and methods used in constructing a commercial building project. We will also learn some of the terminology used in building trades. The chapters are organized according to the five categories in the plans and will give information on each division in the specifications that is associated with that category of the drawings.

Each chapter in this section will give us a look at specific building elements used for that phase of the projects construction and provide examples and illustrations of what those elements look like and how they are used. Study of these examples and illustrations will prepare us for practicing the method for finding construction information in the working drawings and specifications that we learned in Chapter 5.

After a study of each chapter, we will practice the Chapter 5 methods by using an actual set of blueprints and specifications for a commercial building to locate information asked for on chapter worksheets. These exercises will provide practice in the methods learned in Section I, so we become proficient in their use and reduce our search time in the construction documents. It will also reinforce our understanding of the building materials and symbols used on a project and the basic understanding of a commercial construction project.

6

ARCHITECTURAL CATEGORY OF THE DRAWINGS

OVERVIEW

In this unit we will look at the types of information provided in the Architectural category of the Working Drawings. Typical information from each of the Related Specification Divisions will be presented, along with sample illustrations of how that information typically will be shown in the plans, plus examples of symbols used for the construction materials. The Architectural category will provide us the best look at the overall building assembly and materials used in the construction process. A quick review of Figure 6.1 shows some of the basic building materials that will be indicated in the Architectural category of the drawings and how they are typically depicted on the drawings sheets.

DIVISION 1, "GENERAL REQUIREMENTS"

The information contained in Division 1, "General Requirements," pertains to all of the sheets in each category of the working drawings. The requirements in this division govern all of the project scope of work shown in the drawings. Usually, this division will contain information regarding management or administration of the project, and although most of this information cannot be graphically displayed on the drawings, some of it can be shown.

Print and Specifications Reading for Construction, Updated Edition. Ron Russell.
© 2024 John Wiley & Sons, Inc. Published 2024 by John Wiley & Sons, Inc.
Companion website: www.wiley.com/go/printspecreadingupdatededition

RELATED SPECIFICATION DIVISIONS

Division 1, "General Requirements"

Division 4, "Masonry"

Division 5, "Metals"

Division 6, "Woods, Plastics, and Composites"

Division 7, "Thermal and Moisture Protection"

Division 8, "Openings"

Division 9, "Finishes"

Division 10, "Specialties"

Division 11, "Equipment"

Division 12, "Furnishings"

Division 14, "Conveying Equipment"

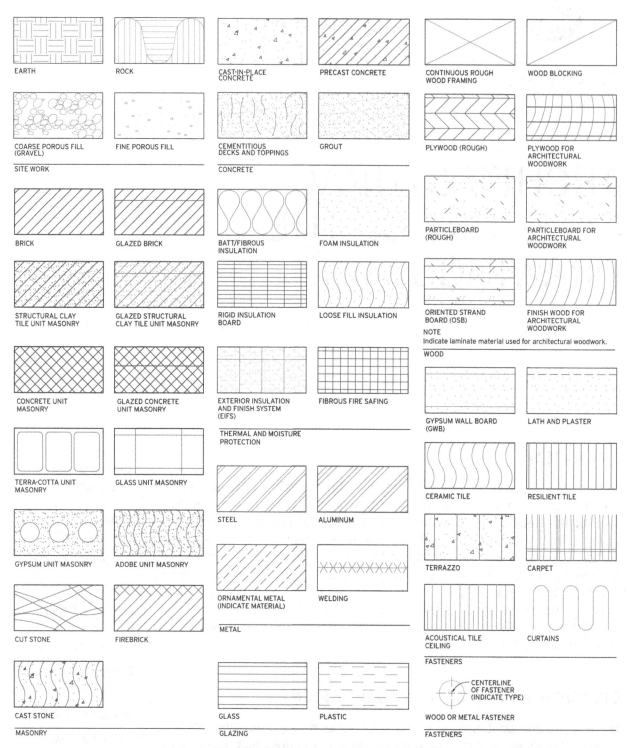

FIGURE 6.1 Basic Building Materials

The sections of Division 1 that could have information shown in the drawings include temporary facilities and controls, materials and equipment, and starting of systems. The temporary facilities and controls information will be typically shown on the civil drawings so the contractor will know where to access utilities and possible locations for setting up temporary offices for use during construction. That would also be the location for identifying information on any site controls necessary for coordinating and controlling access to the construction site. Materials and equipment information could be shown in any of the working drawing categories where materials for the project have special handling requirements or special construction is needed to temporarily protect materials or equipment in place from damage until the project is completed. Starting of systems would outline procedures to follow when beginning the use of certain materials or equipment for completion of the project and will be shown in the working drawing category where the systems are defined. Typically, the information relating to this information will be displayed in the plans in notes for each particular category. These notes should be read before proceeding further to ensure that, if any drawings exist that provide graphical details, the reader can go to those first to be prepared to take any action necessary on the work.

The management or administrative information will typically include subjects such as a summary of work, coordination of project meetings, how submittals will be processed, quality control of the construction, and contract closeout. Most of this information outlines procedures to be followed to accomplish the project work and will not be graphically shown. Therefore, we will not look at Division 1 again as we learn about the other categories of the working drawings.

DIVISION 4, "MASONRY"

Most masonry work that would be included on a commercial project would typically be either structural or used as a finish. In the architectural category, we will only see the masonry work used as a finish. We will look at structural masonry in Chapter 7.

Many commercial construction projects have masonry in the form of brick or clay tile used for an exterior finish. This is obviously a low-maintenance finish material, and affords the owner and architect the opportunities to be creative in the building's exterior design because bricks come in many varied colors and sizes. Figure 6.2 shows some of the symbols for clay and concrete masonry units and sizes they are available in. Also note in this example that brick is assembled in a pattern that is called a *bond*. There are various bonds that can be used to achieve different aesthetic presentations on the exterior of a building. The type of bond pattern used on

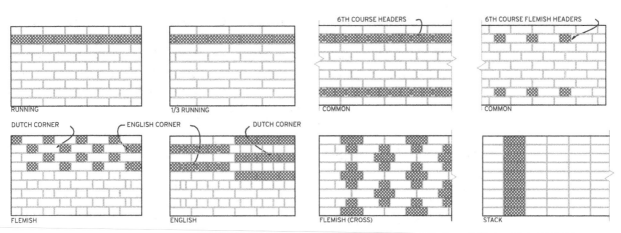

FIGURE 6.2 Typical Brick Patterns

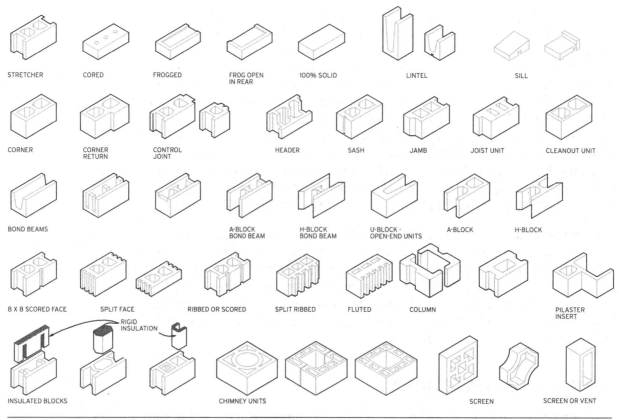

STRETCHER CORED FROGGED FROG OPEN IN REAR 100% SOLID LINTEL SILL

CORNER CORNER RETURN CONTROL JOINT HEADER SASH JAMB JOIST UNIT CLEANOUT UNIT

BOND BEAMS A-BLOCK BOND BEAM H-BLOCK BOND BEAM U-BLOCK - OPEN-END UNITS A-BLOCK H-BLOCK

8 X 8 SCORED FACE SPLIT FACE RIBBED OR SCORED SPLIT RIBBED FLUTED COLUMN PILASTER INSERT

RIGID INSULATION
INSULATED BLOCKS CHIMNEY UNITS SCREEN SCREEN OR VENT

TYPICAL CONCRETE MASONRY UNIT SHAPES

FIGURE 6.3 Typical Concrete Masonry Unit Shapes

the exterior of a project building will typically most easily be seen on the elevation drawings in the architectural category.

Masonry units can also be used on the interior of the building for finishes, and also for partition walls where 4-inch-wide concrete blocks are used to construct walls instead of steel frame and gypsum board. When used as a veneer on an interior wall, the masonry units, typically brick, will be placed similarly to its use on the exterior. When looking at the drawings, the observer would start with a plan view and look for section view call-outs to get more detailed information.

Masonry blocks, or concrete blocks, are used as shown in Figure 6.3 to create partition walls to divide up space instead of conventional materials such as metal studs and gypsum board. This type of masonry construction will typically only go to the bottom of the roof deck or to just above the ceiling height and is not considered a load-bearing wall. Unlike the brick walls, these types of walls will require some type of finishing.

DIVISION 5, "METALS"

Most metals used on a commercial building project would be involved with the structure. However, in the architectural category, miscellaneous metals will be used in various locations, many of which are decorative. Figure 6.4

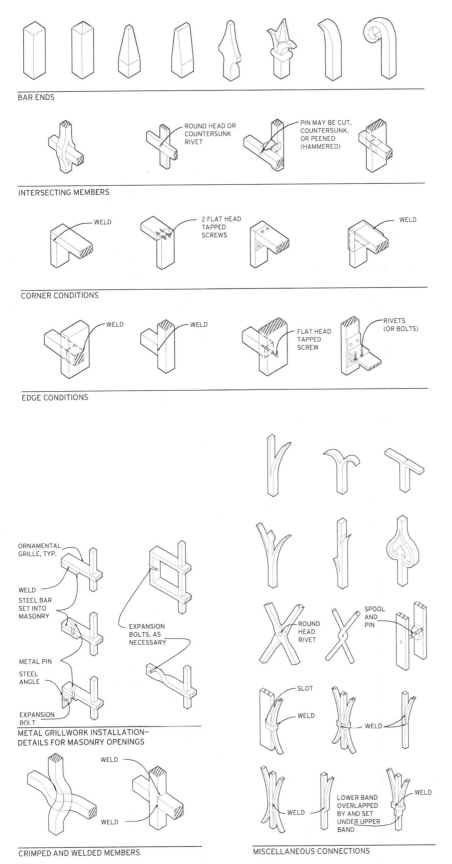

BAR ENDS

INTERSECTING MEMBERS

ROUND HEAD OR COUNTERSUNK RIVET

PIN MAY BE CUT, COUNTERSUNK, OR PEENED (HAMMERED)

CORNER CONDITIONS

WELD

2 FLAT HEAD TAPPED SCREWS

WELD

EDGE CONDITIONS

WELD

WELD

FLAT HEAD TAPPED SCREW

RIVETS (OR BOLTS)

ORNAMENTAL GRILLE, TYP.

WELD

STEEL BAR SET INTO MASONRY

EXPANSION BOLTS, AS NECESSARY

METAL PIN

STEEL ANGLE

EXPANSION BOLT

METAL GRILLWORK INSTALLATION– DETAILS FOR MASONRY OPENINGS

WELD

WELD

CRIMPED AND WELDED MEMBERS

ROUND HEAD RIVET

SPOOL AND PIN

SLOT

WELD

WELD

WELD

LOWER BAND OVERLAPPED BY AND SET UNDER UPPER BAND

WELD

MISCELLANEOUS CONNECTIONS

FIGURE 6.4 Ornamental Metals

shows some of the patterns of metal that could be used on hand rails and partitions, for example.

Another use of metals would be the ladder that would be used at an access point to the roof or a mezzanine floor level. Another typical application of metals would be with the wall flashings such as the roof scuppers, which allow water to escape from the roof.

Other roofing applications for metals would be in the form of flashings and copings. Figure 6.5 show typical applications for these metals, the function of which will be discussed later under Division 7, "Thermal and Moisture Protection."

Another application for metals in a commercial building is as blocking in walls or ceilings where something is needed that is stronger than the partition wall materials to support items that would be hung from or fastened to them. Blocking metals would be mounted behind the finish materials so they would not be seen. The item would be hung or mounted with mechanical fasteners of different types that would be tapped into the metal. The metal framing materials themselves will be discussed later in this chapter under Division 9, "Finishes."

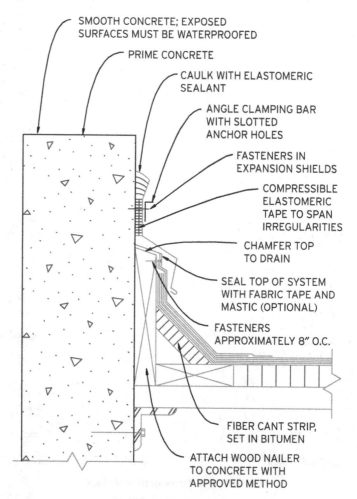

FIGURE 6.5 Flashings

Many of the metals found in this category will be shown in the details that are derived from the plan views, sections, and details.

DIVISION 6, "WOODS, PLASTICS, AND COMPOSITES"

As with masonry and metal materials, the applications for woods and plastics would usually be for major structural elements, which will be discussed in Chapter 7. However, there are still uses for woods and plastics in the architectural category of the drawings.

Because most modern commercial buildings are made of concrete, metals and masonry, woods have been relegated to uses for blocking in a similar manner, as discussed with metals and shown in Figure 6.5, or as trim material at wall junctions at the floor and around doors and windows. Many trim pieces for wall decoration fill a practical application for concealing irregularities in the construction where two surfaces come together, or to protect the wall's surfaces, as with a chair rail. However, this trim application can also be used to enhance, or upgrade the appearance of a room, especially when used in combinations as shown in Figure 6.6. This use will typically be more prevalent in buildings where people traffic is lower and where aesthetics contribute to the business being conducted within the facility, such as with banks or churches. Special pieces of wood are used as trim around doors and windows to achieve distinctive appearances. Otherwise, the materials used might be plastics or vinyls, as with the cove base trim around the floor. These materials will be usually be indicated on the drawings in section views of the walls or in the details for the wall partitions.

The other primary application for woods and plastics in a modern commercial building will be in the millwork, or cabinetry. Most cabinetry is shown in an elevation with only a general outline of its overall size and shape and door swings and mounting configurations indicated. The specifications will include the primary definition provided by the architect of the construction of the cabinetry. The architect will not provide a great deal of detail on the cabinetry construction; the general contractor will

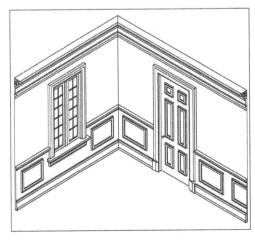 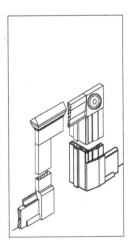

FIGURE 6.6 Wood Applications

include those details on the shop drawings submitted for review. There we will see the main applications of the woods used, as well as the primary application for the plastic used in the form of plastic laminates.

Plastic laminates consist of different layers of materials impregnated with plastics and cured under heat and pressure. They are applied to wood products with adhesives to provide durable and washable surfaces, as shown in Figure 6.7 on countertops. These materials can also be used for

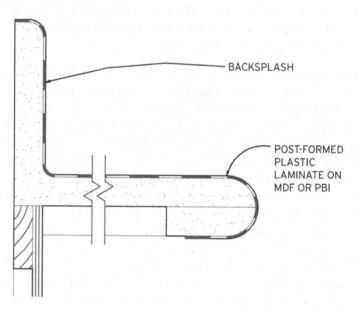

BACKSPLASH

POST-FORMED PLASTIC LAMINATE ON MDF OR PBI

POST-FORMED LAMINATE COUNTERTOP

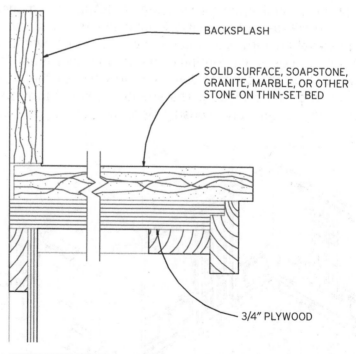

BACKSPLASH

SOLID SURFACE, SOAPSTONE, GRANITE, MARBLE, OR OTHER STONE ON THIN-SET BED

3/4" PLYWOOD

STONE COUNTERTOP

FIGURE 6.7 Countertops

wall coverings, flooring, and on doors. There are many different types of colors and textures. Though only noted in the drawings, the main definition of these plastic materials will be in the specifications.

DIVISION 7, "THERMAL AND MOISTURE PROTECTION"

There are many thermal and moisture applications on a commercial building. We will examine the primary applications, roofing, insulation, and sealants. We will look at roofing systems first.

There are three primary roofing systems used on commercial buildings: built-up roofs, membrane roofs, and metal roofs. It is noteworthy that the roof slope is established by the structure, not the roofing materials, which are only applied to the structural surface. Most commercial buildings have what are called *flat roofs,* with minimal slope to allow drainage. The roofing system used by the architect will take into consideration the roof configuration and building design. This configuration will be shown on a plan view, roof plan and will allow the observer to see the general slope of the roof indicated with directional arrows, and the build-up areas used to facilitate drainage. This plan will also indicate points of drainage, or where water is planned for exiting the roof.

Built-up roofing systems consist of multiple plies of roofing felts applied in layers over the roof deck or insulation board with some type of ballast, or aggregate, to provide weight and to break up live loads from the wind, rain, and snow. Figure 6.8 shows an example of the materials used for this

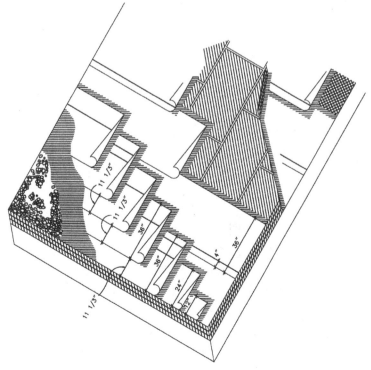

FIGURE 6.8 Built-up Roof

element of the building. You will not usually see this detail in the working drawings since the roofing specifications will describe the required materials' quality and a specific manufacturer, who will be responsible for determining the application methods. Usually, since the manufacturer is providing a significant warranty, an installer approved by that manufacturer accomplishes the roofing system installation. The details that will be shown in the drawings will be like Figure 6.9 for flashings and terminations of roofing materials.

The built-up roofing system will require that the roofing materials be terminated so that moisture and air cannot penetrate the building envelope. Typical flashings include the base flashings, counter flashings, and cap flashings, as shown in Figure 6.9, where a parapet wall is the termination point. The roofing felts will start up the parapet wall surface. Base flashings will be attached to the wall surface and continue down the wall covering the end of the roofing felts. A counter flashing system will be designed using a cap flashing to cover the connection points of the base flashings. If the base flashings and counter flashings terminate at the top of a parapet wall, a final trim cap called coping, usually of a colored metal, will be used to cover the final attachment locations. The counter flashing is shaped so that the coping snaps into place.

Another significant element of a built-up roofing system is the gravel guard shown in Figure 6.10. This element prevents the ballast, or aggregate, from washing off the building and exposing the roofing materials. Roof penetrations are also significant and are typically flashed as shown in Figure 6.11. Another option for penetrations is a device called a pitch

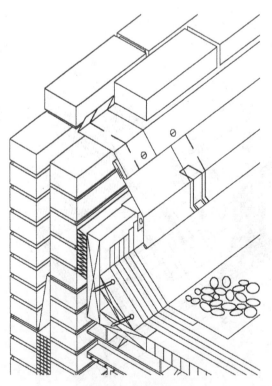

FIGURE 6.9 Flashings

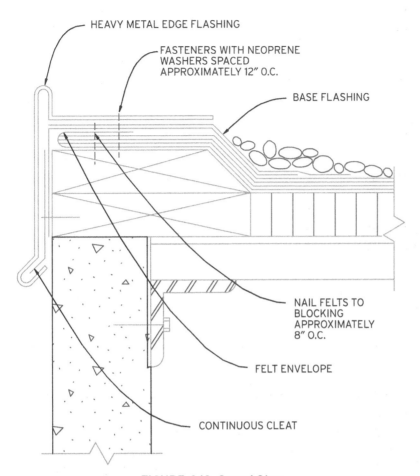

FIGURE 6.10 Gravel Stop

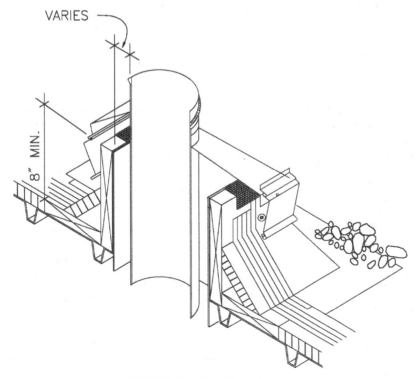

FIGURE 6.11 Roof Penetration

pan. A pitch pan is placed around a pipe or conduit as it penetrates the roof. Then the pan is filled with a sealant, such as tar pitch, to seal the roof penetration. Another typical penetration is for equipment such as HVAC units. Since these units are too large for a pitch pan, a curb is used to set the equipment on and the flashing system is used to seal around the curb.

Membrane roofing systems consist of a single-ply membrane of some type of elastomeric material that is laid on the roof decking materials, with the joints closed using a sealant that is heated to fuse the joints and a ballast applied. The flashing techniques are similar to those of the built-up roof, except that a single ply of the same material as the roof is used for flashings and roof drains. Ballast is an option with most single-ply roofing systems. Expansion joints for single-ply membrane roofs allows the roofs of individual buildings to expand and contract with the weather without creating gaps or holes.

The most common metal roof is the standing seam roof, and although this type of roofing system is durable, it is more subject to damage from the elements. This type of system tends to be more expensive; however, it can allow the architect and owner to do more with the exterior aesthetics of the building. Various types of standing seams allow the metal roofing panels to be interlocked to cover the roof decking materials and provide a moisture barrier for the building. These seams, which stand above the roofing surface, are where this type of system gets its name. Figure 6.12 shows a typical detail of how the standing seam roofing materials are formed for applying to the building. Unless it is designed as part of the architectural character of the building, this system is not often used.

The roofing system provides both of the characteristics in Division 7, "Thermal and Moisture Protection." We have looked at the moisture protection aspect; now let's briefly talk about the thermal aspect.

Most commercial buildings will incorporate some type of insulation in the roofing system for economy reasons alone, not to mention occupant comfort. However, some buildings (such as warehouses) where the number of occupants is small and their work does not require dexterity might have minimal insulation, if any. With the roofing system, a type of rigid board with thermal qualities will be adhered to the roof deck, either through mechanical means or using adhesives. Then, the roofing materials will be placed over the insulation board. However, this insulation board can be omitted and the roofing materials placed directly on the deck. The insulating qualities will be specified according to the building construction type, location of the building, and the building's use. This same rigid board could also be used on other surfaces such as flooring and on vertical walls, where it would not be convenient to use batt or loose insulation materials. Insulating materials will have an "R" value, which is used to measure their thermal resistance to the flow of heat and is selected by the architect to match the building's requirements.

Batt or loose insulation materials are used where there are irregular spaces that can contain the materials. Figure 6.13 compares the various types of insulation. Batt materials are used to fill the spaces in walls and

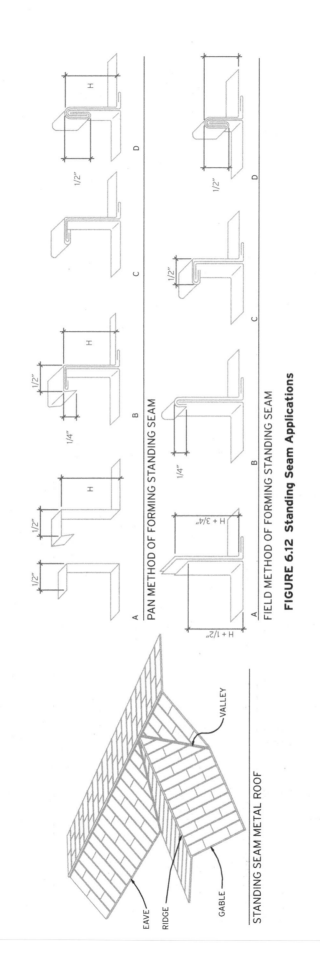

A   B   C   D

PAN METHOD OF FORMING STANDING SEAM

A   B   C   D

FIELD METHOD OF FORMING STANDING SEAM

FIGURE 6.12 Standing Seam Applications

STANDING SEAM METAL ROOF

EAVE

RIDGE

GABLE

VALLEY

MATERIAL PROPERTIES OF COMMON BUILDING INSULATION

BUILDING INSULATION	DENSITY (LB/CU FT)	RESISTANCE @ (HR/SQ FT °F / BTU PER IN. THICKNESS)	WATER VAPOR PERMEABILITY	WATER ABSORPTION (% BY WEIGHT)	SURFACE BURNING CHARACTERISTICS		TOXICITY	EFFECTS OF AGING	TEMPERATURE	DEGRADATION DUE TO			CORROSIVENESS
					FLAME SPREAD	SMOKE DEVELOPED		DIMENSIONAL STABILITY		MOISTURE	FUNGAL OR BACTERIAL GROWTH	WEATHERING	
Glass fiber batts and blankets rigid boards	1.5-4.0 / 4.0-9.0	3.14 / 3.8-4.3	100 / 100	2% / 10%	15-25 / 0-25	0-50 / 0-50	Some fumes if burned	None	OK below 450xF	None	None	None	None
Rock or slag wool	1.5-2.5	2.9-3.7	100	2%	0-25	0-20	None	None	600°F	Transient	None	None	None
Cellulose (loose blown)	2.0-3.0	2.8-3.7	100	15%	0-50	0-45	CO if burned	Settles 0-20%	Possible with long exposure	Possible with long exposure	Maybe	Possible with long exposure	Steel, aluminum, and copper
Molded polystyrene (rigid boards)	0.9-1.8	3.6-4.4	1.2-5.0	2-3%'	25	10-400	CO if burned	None	If above 165°F	None	None	UV degrades	None
Extruded polystyrene (rigid boards)	1.6-3.0	4.0-6.0	0.3-0.9	1-4%	25	10-400	CO if burned	None	If above 165°F	None	None	UV degrades	None
Polyurethane (rigid boards)	1.7-4.0	$5.8\text{-}6.2^2$	2-3	Negligible	25-75	155-200	CO if burned	0-12% change	If above 250°F	?	None	None	None
Polyisocyanurate (rigid boards)	1.7-4.0	$5.8\text{-}7.8^2$	2.5-3.0	Negligible	25	55-200	CO if burned	0-12% change	If above 250°F	?	None	None	None
Perlite (loose fill)	5-8	2.63	100	Low	0	0	None	Settles 0-10%	If above 1200°F	None	None	None	None
Vermiculite (loose fill)	4-10	2.4-3.0	100	None	0	0	None	Settles 0-10%	If above 1000°F	None	None	None	None
Phenolic (foamed-in-place or rigid boards)	2.5-4.0	4.4-8.2	1.0 for rigid	1-2% for rigid	20-50	0-35	None	None	If above 250°F	None	None	UV degrades	None

FIGURE 6.13 TYPES OF INSULATION

under floors and are typically shown as depicted in these examples. Batt insulation comes in various thicknesses and widths to accommodate the typical framing of a building. This type of insulation holds its shape well and can be put in place and covered. Loose insulation must be placed in the space after it is completed so that it will take the shape of the space. Loose wet applied insulation can be used where no definable space is available. This material can be sprayed directly onto the exposed surface and allowed to dry. An external insulation and finish system is often used on the exterior of masonry buildings to provide insulation and finish for the appearance of the building (EIFS).

Another aspect of moisture proofing involves elements of the building that are below grade. Figure 6.14 shows a below-grade wall where water-resistant membranes have been used to prevent moisture from the soils to penetrate the concrete. This moisture must not be allowed to penetrate the concrete or else valuable salts and Portland cement will be leached out of the wall and its structural integrity will be compromised. This moisture-proofing effect could also be accomplished using some type of mastic material applied directly to the wall surface before backfill is placed against it. The same moisture protection considerations must be made when dealing with a slab-on-grade. Notice in Figure 6.14 that the slab has a moisture barrier, or membrane, in place underneath it and a rubber water stop is used where the slab meets the perimeter walls.

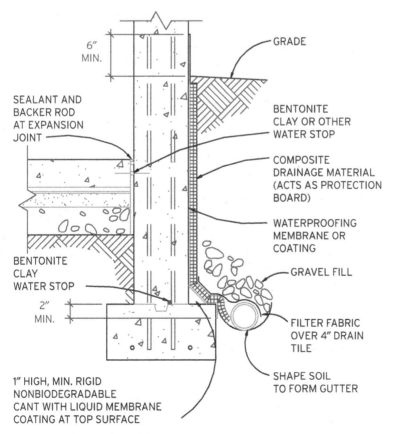

FIGURE 6.14 Below-grade Waterproofing

One last consideration in moisture proofing would be the use of sealants or caulk to seal joints or gaps between materials that require some element of separation, typically for expansion and contraction. Masonry joints and building joints typically require some type of sealant to close the gap and protect the interior of the building. The material is applied directly to the surfaces if the space or gap is small and does not require a significant amount of material. However, a backer material is needed for the gap if the space is larger and the sealant material would not be able to retain its shape if too much material is applied to the joint.

DIVISION 8, "OPENINGS"

Doors, windows, and glass complete the enclosure of the building while allowing passage in and out of the facility. These closures have many different sizes and shapes, depending on their function and location in the building. To understand the uses of doors by the architect, let's look at the basic elements of these materials.

Doors

In Figure 6.15, we can see the basic elements of a door. If we consider a rough opening in a wall, the first element of the door to be installed would be the side and top jambs. These elements would be installed level and plumb to ensure that the door swings square. Then casing, or trim, would be attached to the jambs and would extend past the finish materials in order to cover the irregularities between the jambs and the rough opening. Then another trim piece would be installed inside the jambs as a stop, which is used to prevent the door from swinging past the resting place of the door when in the closed position. This is a simplified description of the door elements, but it provides enough information so that we can see these elements later when we discuss how they are applied in today's commercial building. The other elements of the door include the door itself, plus any hardware necessary for the proper function of the door.

There are two basic ways to identify the types of doors used on a building, by operation and by appearance or construction. When we say by operation, we are indicating how the door functions—that is, swinging, bypass or surface sliding, or bifolding. Figure 6.16 shows various types of doors by their appearance, which also sometimes facilitates their function, such as shown by the louvered door to provide air flow and the glass doors to allow light into a building space. Roll-up doors and grilles are used for crowd control in buildings and along with overhead doors to close off large spaces, as discussed later.

Proficiency in identifying these basic door elements and types is necessary for understanding how they are applied in today's commercial building. The elements that we have discussed are as they would be if they were assembled on site out of traditional materials such as wood. However, in today's commercial building, most of these materials would not hold up to building traffic and use. Today, most of these door framing elements are

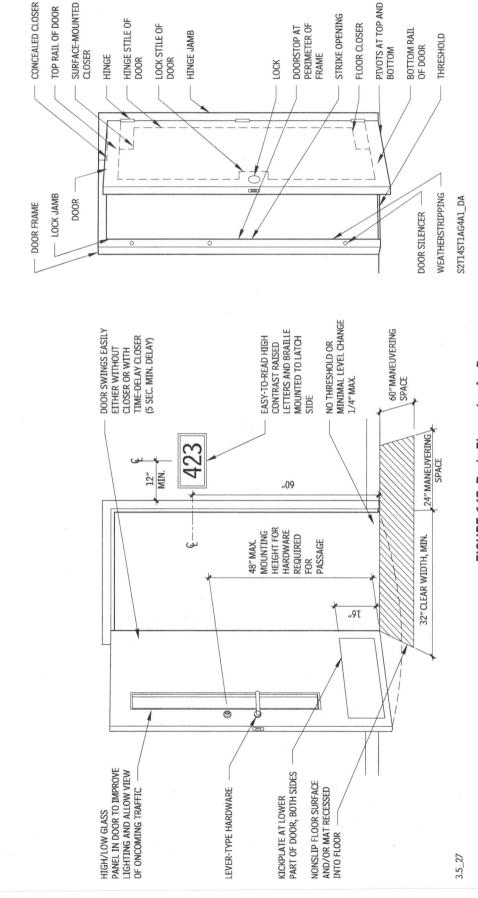

CONCEALED CLOSER
TOP RAIL OF DOOR
SURFACE-MOUNTED CLOSER
HINGE
HINGE STILE OF DOOR
LOCK STILE OF DOOR
HINGE JAMB
LOCK
DOORSTOP AT PERIMETER OF FRAME
STRIKE OPENING
FLOOR CLOSER
PIVOTS AT TOP AND BOTTOM
BOTTOM RAIL OF DOOR
THRESHOLD

DOOR FRAME
LOCK JAMB
DOOR

DOOR SILENCER
WEATHERSTRIPPING
S2T14ST1AG4A1_DA

HIGH/LOW GLASS PANEL IN DOOR TO IMPROVE LIGHTING AND ALLOW VIEW OF ONCOMING TRAFFIC

DOOR SWINGS EASILY EITHER WITHOUT CLOSER OR WITH TIME-DELAY CLOSER (5 SEC. MIN. DELAY)

EASY-TO-READ HIGH CONTRAST RAISED LETTERS AND BRAILLE MOUNTED TO LATCH SIDE

423

12" MIN.

60"

NO THRESHOLD OR MINIMAL LEVEL CHANGE 1/4" MAX.

60" MANEUVERING SPACE

24" MANEUVERING SPACE

48" MAX. MOUNTING HEIGHT FOR HARDWARE REQUIRED FOR PASSAGE

32" CLEAR WIDTH, MIN.

16"

LEVER-TYPE HARDWARE

KICKPLATE AT LOWER PART OF DOOR, BOTH SIDES

NONSLIP FLOOR SURFACE AND/OR MAT RECESSED INTO FLOOR

3.5_27

FIGURE 6.15 Basic Elements of a Doorway

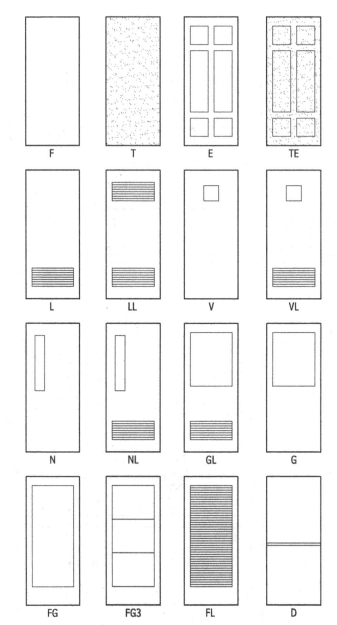

FIGURE 6.16 Doors Appearance

made into one unit called hollow metal frames. Figure 6.17 shows typical metal frames that incorporate all the elements—including the jambs, stops, and trim pieces—in one unit. These metal frames are installed as a unit in the building before the finish materials are applied and are secured into the rough openings, as shown in Figure 6.18. Figure 6.19 shows the components of a metal door frame.

These types of frames require more planning than with wood door elements. With wood door elements, the hardware installation can be field fitted to the wood members. On the metal frames, the hardware assembly locations must be premilled into the frames. That is where we must carefully consider door swing and the application of the hardware. Most door hardware will be installed in typical locations, as shown in Figure 6.20.

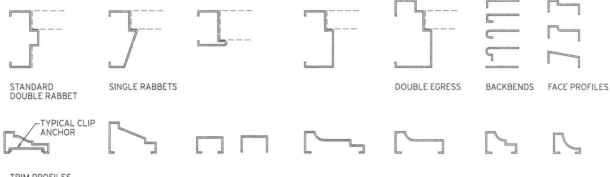

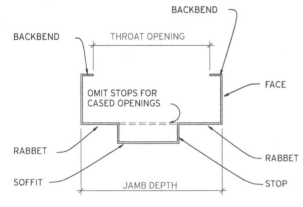

STANDARD DOUBLE RABBET

NOTE

Maximum gauge is 10; consult manufacturers for lighter gauges.

VARIOUS STANDARD PROFILES

	JAMB DEPTH (IN.)								
	2¾	3	3¾	4¾	5½	5¾	6¾	7¾	8¾
Rabbet*	Single rabbet only			1¹⁵⁄₁₆ in. standard for 1¾ in. door					
Soffit*									
Rabbet*				1⁹⁄₁₆ in. standard for 1³⁄₈ in. door					
Backbend	½	⁷⁄₁₆	½	½	¾	½	½	½	½
Throat	1¾	2¹⁄₈	2	3¾	4	4¾	5¾	6¾	7¾

* Omit stops for cased opening frames.

FIGURE 6.17 Hollow Metal Door Frames

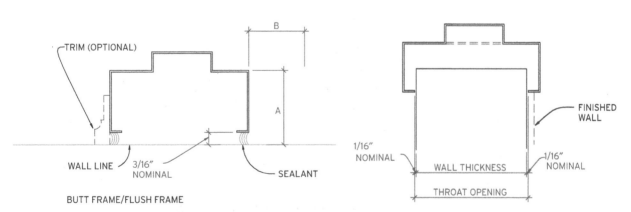

FIGURE 6.18 Door Frame Openings

Once the order is placed for the metal frames, the manufacturer will make the mounting locations in the finished frames based on this information in the drawings. This is where it is important that the contractor ensures that he understands what hardware is required for a door, and what direction it must swing.

If you look at Figure 6.21, you can see the method used to determine the swing conventions of doors. This is important to note because the

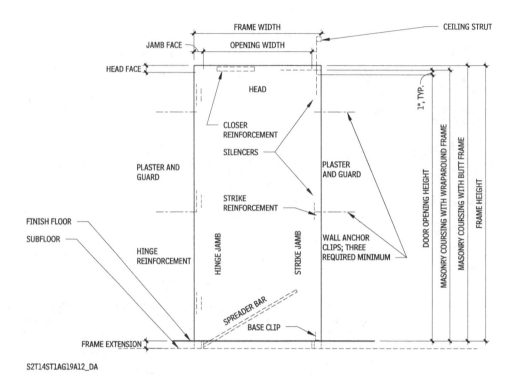

S2T14ST1AG19A12_DA

FIGURE 6.19 Metal Door Frame Components

DOOR HARDWARE

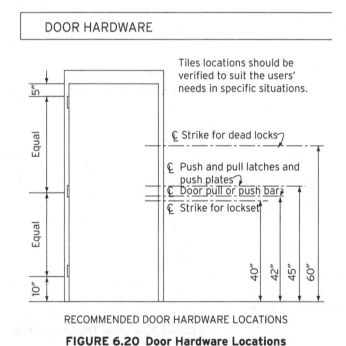

Tiles locations should be verified to suit the users' needs in specific situations.

RECOMMENDED DOOR HARDWARE LOCATIONS

FIGURE 6.20 Door Hardware Locations

manufacturer of the frames will be placing the door hardware mounts based on the direction of swing indicated. To avoid confusion from various methods for identifying door swing directions, many manufacturers have developed their own methods and will provide this to the contractor for ordering. The architect determines the swing of the doors considering codes and the planned use of the building. These door swings cannot usually be changed. If the contractor does not describe the swing of a

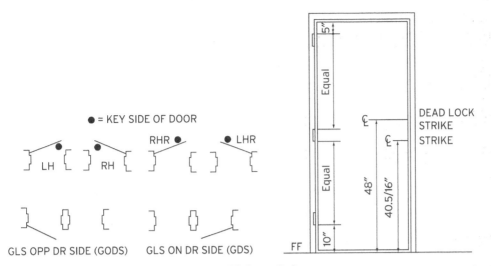

FIGURE 6.21 Door Swings

door correctly, the manufacturer will place the hardware mounts in the wrong locations and the metal frame cannot be used. If the door elements are made of wood, you might be able to salvage the materials if a mistake is made.

As you can see, a lot of information is required for doors. The plan views would be consumed with this information if it were all to be placed there. Instead, the architect will use some type of numbering or letter system to identify each door and use a schedule to provide this information. In Figure 6.22, look at door number A113 in the schedule. This door size is 3'–0"×4'–0, "it is a Type "F" door, it is a solid core wood door, and if we were to check the specifications, we would see that it is to have a stained finish, it has an hollow metal frame, and it receives a lockset. As you can see, this would be a significant amount of information to place on the plan view where the door is shown. Having all of this information on the schedule is much more convenient and allows the drawings to remain clear.

Another type of door typically used in most commercial buildings is the overhead door. This is very much like a residential garage door, only heavier to withstand the additional use. These doors can be made of wood, but typically are steel, aluminum, or fiberglass. These doors can be of multiple sizes and could be up to 20 feet high and up to 30 feet wide to fit commercial applications. If not manually operated, they can be actuated by a chain hoist or electric motor. These doors are generally used at ramp entrances, dock areas, vehicle entrances, or other places where a significant amount of bulk traffic is expected. These door installations would be typically shown in the architectural category of the drawings.

Windows

For windows, the same basic principles apply that we used with doors. First, we should look at the basic elements as shown in Figure 6.23. Again, we see the use of jambs and casing or trim materials, and other elements,

GENERAL DOOR AND WINDOW NOTES:

1. REFER SHEET AP 6.01 FOR DOOR SCHEDULE.

2. REFER TO SHEET AP 5.01 FOR LINTEL SCHEDULES AND DOOR AND WINDOW JAMB STUD FRAMING SCHEDULE.

3. ALL DOORS TO BE 1 3/4" THICK (UON)

4. ALL STL FRAMES TO HAVE A 2" FACE DIMENSION UON, W/ 5/8" STOP. 3/4" STOP REQUIRED FOR FIRE RATED FRAMES.

5. STEEL FRAME DEPTHS INDICATED IN THE DETAILS ARE NOMINAL. PROVIDE STANDARD WIDTH FRAME WITH THROAT OPENING APPROPRIATE TO WALL TYPE.

6. COORDINATE WITH MECH DOCUMENTS FOR DOOR UNDERCUTS AND DOOR LOUVERS, IF ANY.

7. ALL LIGHTS IN DOORS TO HAVE SAFETY GLAZING (UON).

8. EXTERIOR DOORS TO HAVE TINTED GLAZING (UON) AND INTERIOR DOORS TO HAVE CLEAR GLAZING.

9. FIELD VERIFY DIMENSIONS PRIOR TO FABRICATING FRAMES.

10. PROVIDE WD BLKG IN GYP BD WALLS BEHIND DOORS SCHEDULED TO HAVE WALL MOUNTED DOOR STOPS. (REFER HARDWARE SCHEDULE SUBMITTAL).

11. REFER TO DTLS B5 AND B6/AP 5.01 FOR TYP WINDOW AND DOOR CONTROL JOINTS IN MASONRY WALLS. REFER DTLS- B5 AND B6/AP5.02 FOR TYP WINDOW AND DOOR CONTROL JOINTS IN STEEL STUD FRAMED WALLS.

12. FLOOR MATERIAL CHANGES BETWEEN ROOMS TO OCCUR UNDER DOOR.

13. AT ALL EXISTING STL FRAMES WHERE DOOR SWING IS TO BE CHANGED, PATCH, REPAIR AND GRIND SMOOTH EXISTING STL DOOR FRAME. PRIME AND USE ONLY IF FRAME IS TO BE RETAINED. PAINT ACCORDING TO COLOR SCHEDULE SHEET AF 6.01.

DOOR SCHEDULE REMARKS

1. PAIR OF 3'-0"W x 7'-0"H DOORS
2. REVERSE EXISTING DOOR SWING
3. PROVIDE NEW DOOR AT EXISTING WINDOW LOCATION. REFER DEMOLITION PLAN FOR ADDITIONAL INFORMATION.

DOOR SCHEDULE

| DOOR NUMBER | DOOR | | | | FRAME | | | | | FIRE RATING | REMARKS |
	WIDTH x HEIGHT	TYPE	MATERIAL	GLAZING	TYPE	MATERIAL	HEAD	JAMB	SILL		
A101	3'-0" x 4'-0"	FGIR	ALUM	SAFETY	3	ALUM STFT	E5/AP4.02	C5/AP4.02	B6/AP4.02	–	–
A102	3'-0" x 4'-0"	FGIR	ALUM	SAFETY	4	ALUM STFT	E1/AP4.02	E2/AP4.02	B6/AP4.02	–	–
A103	2 @ 3'-0" x 4-0	FGIR	ALUM	SAFETY	6	ALUM STFT	E1/AP4.02	E2/AP4.02	B6/AP4.02	–	1
A104	2 @ 3'-0" x 4-0	FGIR	ALUM	SAFETY	5	ALUM STFT	D5/AP4.02	C5/AP4.02	B5/AP4.02	–	1
A112	3'-0" x 4'-0"	F	WOOD	–	1	STL	E6/AP4.02	D6/AP4.02	–	–	–
A113	3'-0" x 4'-0"	F	WOOD	–	1	STL	E6/AP4.02	D6/AP4.02	–	–	–
A114	3'-8" x 4'-0"	F	STL	–	2	STL	E3/AP4.02	D3/AP4.02	–	–	–
A115A	3'-0" x 4'-0"	F	STL	–	1	STL	D4/AP4.02	C4/AP4.02	–	–	3
A115B	3'-0" x 7'-0"	EXISTING	EXISTING	EXISTING	EXISTING	EXISTING	EXISTING	EXISTING	EXISTING	–	2
A116A	3'-0" x 4'-0"	F	STL	–	1	STL	D4/AP4.02	C4/AP4.02	–	–	3
A116B	3'-0" x 7'-0"	EXISTING	EXISTING	EXISTING	EXISTING	EXISTING	EXISTING	EXISTING	EXISTING	–	2

FIGURE 6.22 Door Schedule Example

such as the rails used to secure the glass, and the sill pieces for interior and exterior applications. All of these materials are installed square into the rough opening. Figure 6.24 shows use some of the types of windows available by type or function.

In most commercial buildings, the windows today are fixed, or inoperable, since the inside environment is controlled for air conditioning purposes. The windows are also typically purchased and installed as a unit, again to eliminate the need for constructing each element in place. The unit is

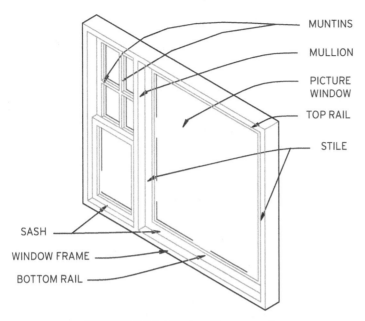

FIGURE 6.23 Window Components

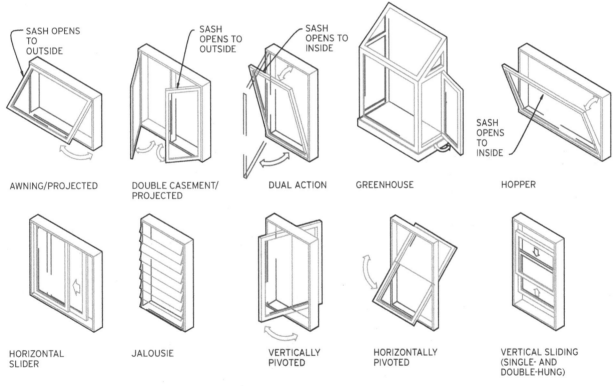

FIGURE 6.24 Windows by Operation

typically hollow metal, as with the doors, and is positioned into the rough opening before the finish materials are applied. As with the doors, the windows will be numbered or identified with a letter that corresponds with a schedule that allows us to minimize the amount of information that must be placed on the plan views.

Storefront Assemblies

On many commercial buildings, the door and window elements are combined to create what are called *storefront* assemblies (see Figure 6.25). These assemblies are typically made out of extruded materials, such as aluminum, and assembled on site into the rough opening, usually at entrances or foyers. The rough opening dimensions are provided to the installer, who will cut the material pieces on site to fit the opening. The drawings will typically only show the basic dimensions and elements in an elevation view.

Glass and glazing, which are the materials and systems used to hold the glass in place, will be primarily defined in the specifications and only shown minimally in the drawings.

DIVISION 9, "FINISHES"

Finishes consist of many different types of materials on surfaces that are both interior and exterior. Each finish material will be assigned some type of code, as shown in the Materials and Color Summary in Figure 6.26.

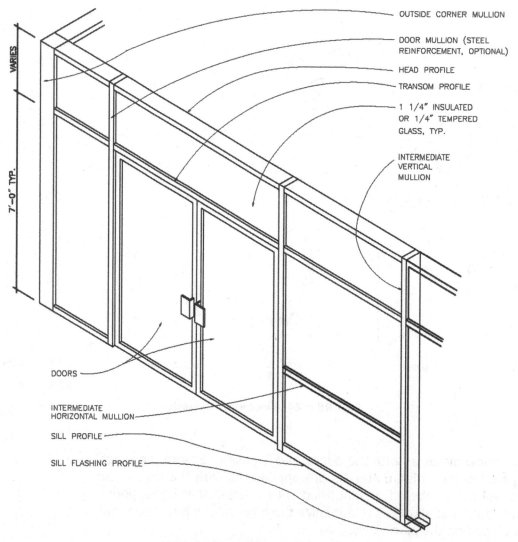

FIGURE 6.25 Typical Storefront

MATERIAL AND COLOR SELECTION SUMMARY

Elementary School

PROJECT NO.   REVISED 06/15/09

Surface/Material	Area	Type	Manufacturer	No.	Color
Floors					
VCT # 1	All Specified	12 x 12	Armstrong	51811	Antique White
VCT # 2	All Specified	12 x 12	Armstrong	57504	Chocolate
VCT # 3	All Specified	12 x 12	Armstrong	51910	Classic Black
CPT-1 Carpet	All Specified	01957 Crayon	C&A	48010	Precious Metal
PT-1 Porcelain Tile	Corridor	12x12/12x24/24x24	Interceramic	Collection-Concrete	Beige
PT-2 Porcelain Tile	Corridor	12x12/12x24/24x24	Interceramic	Collection-Concrete	Brown
PT-3 Porcelain Tile	Corridor	12x12/12x24/24x24	Interceramic	Collection-Concrete	Black
CT-1 Ceramic Tile	Restrooms/Girls	2x2 - Keystones Collection	Daltile	D175	Elemental Tan Speckle
CT-2 CeramicTile	Restrooms/Boys	2x2 - Keystones Collection	Daltile	D202	Uptown Taupe Speckle
CT-3 CeramicTile	Restrooms	2x2 - Keystones Collection	Daltile	D335	Almond
CT-4 Ceramic Tile	Restrooms	2x2 - Keystones Collection	Daltile	D007	Cinnamon Range
CT-5 Ceramic Tile	Restrooms	2x2 - Keystones Collection	Daltile	D311	Black
QT-1 Quarry Tile	Kitchen	6 x 6 Textures	Daltile	OT11	Chocolate
QT-2 Quarry Tile	Kitchen	6 x 6 Textures	Daltile	OTO9	Desert Tan
QT-3 Quarry Tile	Kitchen	Suretread 6 x 6	Daltile	OQ75	Golden Brown
Base					
Base 1	All specified	4" Rubber Base	Roppe	P147	Light Brown
Base 2- Ceramic Tile	Restrooms/Girls	Cove Base 2x2	Daltile	D175	Elemental Tan Speckle
Base 3- Ceramic Tile	Restrooms/Boys	Cove Base 2x2	Daltile	D202	Uptown Taupe Speckle
Base 4- Porcelain Tile	All specified	Cove Base 6x8	MasterTile	A1124	Red Riding Hood
Base 5-Quarry Tile	Kitchen	Cove Base 5x6	Daltile	OT11	Chocolate
FG-1 Floor Grout	All Specified (Corridors)	1/8" Grout line	Custom	122	Linen
FG-2 Floor Grout	All Specified (Restrooms)	1/8" to 1/4" Grout line	Custom	135	Mushroom
FG-3 Floor Grout	Kitchen	1/8" to 1/4" Grout line	Custom	95	Sable Brown
Walls					
Paint # 1 - Field Color	All specified	Paint	Sherwin Williams	SW6126	Navajo White
Paint # 2 - Accent	All specified-Accent	Paint	Sherwin Williams	SW6142	Macadamia
Paint # 3 - Accent	All specified-Accent	Paint	Sherwin Williams	SW6384	Cut the Mustard
PWT-1 Porcelain Tile	Corridors	12 x 12 - Color Blox Collection	MasterTile	A1101	Sandbox
PWT-2 Porcelain Tile	Corridors	12 x 12 - Color Blox Too Collection	MasterTile	A1124	Red Riding Hood
CWT-1 Ceramic Wall Tile	Restrooms	12 x 12 - Ridgeview Collection	Daltile	RD02	Beige
CWT-2 Ceramic Wall Tile	Restrooms/Girls Accent	12 x 12 - Ridgeview Collection	Daltile	RD05	Rust
CWT-3 Ceramic Wall Tile	Restrooms/Boys Accent	12 x 12 - Ridgeview Collection	Daltile	RD06	Blue/Gray
GMT-1 Glass Mosaic Tile	Cafeteria/Accent	1 x 2 Gloss Mosaic	Daltile		Grape Juice
WG-1 Wall Grout	All Specified	1/8" to 1/4" Grout line	Custom	382	Bone
AP-1 Acoustical Panels	All Specified	Acoustical Panels	Interface Fabrics	758	Desert Sand
AP-2 Acoustical Panels	All Specified	Acoustical Panels	Interface Fabrics	298	Medium Grey
Casework					
PLAM #1 - Doors and Cabinets	All Specified	Plastic Laminate	Wilsonart	7936-07	Williamsburg Cherry
PLAM#2 - Countertops	All Specified	Plastic Laminate	Wilsonart	4887-38	Tan Soapstone
Lockers					
L-1 Lockers	Corridors	Metal Lockers	Penco	O12	Tawny Tan
L-2 Lockers	Athletics	Metal Lockers	Penco	736	Burgandy
Partitions					
TP-1 Toilet Partitions	All Specified	Polymer	Global Partitions	2000	Black
Wash Fountains					
Terreon Solid Surface	Restrooms	Acrylic/Resin Solid Surface	Bradley	River	Riverstone

FIGURE 6.26 Material and Color Summary

These codes are used for paints, tiles, and carpets, as well as other materials. This allows a finish schedule to be used on the drawings to avoid cluttering up details with the finish materials' nomenclature. The plan views will have all the rooms numbered for identification. Then, a schedule will be created with each room number, and all the wall, floor, and ceiling finishes for that room will be listed. To determine what finishes are used in a particular room, the observer would look for the room number on the plan view, find that room on the schedule and identify the finish codes, and then look at the finish legend to identify the materials. The specifications would then be used to determine suppliers, grade requirements, and colors for the finish materials.

We have already looked at masonry materials used as interior and exterior finishes, and the use of wood for finish trim, so we will not address those items here. Instead, we will concentrate on those materials that are applied to the basic structure to form the final exposed surfaces that we would expect to be visible in a building. These surfaces include walls, ceilings, and floors. We will discuss walls first.

Walls

Finishes for walls will include the base materials that are used to cover any structural or partition materials, including concrete blocks, bricks, and sheet metal studs. We should recall here that most sheet metal stud walls in a commercial building are called partition walls. Their construction is basically the same in all applications; however, the materials used vary from wall to wall, depending on their location within the building. On the plan view, each wall is called out with a designation of wall type S1 through S8. The drawing reader would note the callout for a wall in question—say, S8—and would turn to the schedule and look at the partition type S8 to determine what materials are used for that particular wall. This eliminates the need to indicate all of these materials on the plan view, which would clutter up the drawing and make it difficult to read.

Most partitions will consist of sheet-metal studs and some type of wallboard or gypsum board, typically called drywall. Drywall is a board with a gypsum core and covered on both sides with paper. It typically comes in sheets 4 feet wide and 8 or 10 feet long. It has preformed beveled edges that facilitate the finishing of its edges to hide the joints where sheets meet, as shown in Figure 6.27. Drywall also comes in various thicknesses and composites of materials. Regular drywall can be obtained in 1/4-inch, 3/8-inch, 1/2-inch, and 5/8-inch thicknesses and is used in various layers to construct typical partition walls. Coreboard is one inch thick and is used for enclosing mechanical chases, elevator shafts, and stairwells. Drywall with an aluminum foil backing is used typically on exterior walls, with the foil acting as a vapor barrier. Water-resistant board is used in high moisture areas, such as restrooms and showers, and serves as a base for ceramic tile applications. Other gypsum boards contain additives such as fiberglass and chemicals to increase the boards' fire resistance. Other types of gypsum board come with sides prefinished with paper or vinyl coverings that are available in different colors and textures.

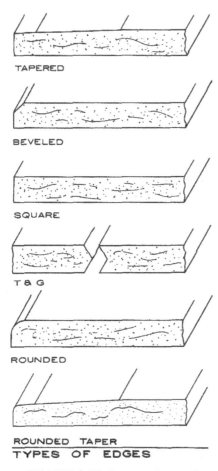

TAPERED

BEVELED

SQUARE

T & G

ROUNDED

ROUNDED TAPER

TYPES OF EDGES

FIGURE 6.27 Gypsum Drywall

In most applications, the gypsum board is applied directly to above-grade masonry or concrete walls, or fastened directly to wood or sheet-metal stud framing. Once applied, the joints between each sheet must be hidden by a process called taping and mudding, or bedding. This process involves applying several layers of paper tape and joint compound to each joint and smoothing them out until the joints are no longer visible. Once the drywall joints are concealed, they can then be textured and painted to complete the finish. Painting is only one method for coating materials for a finish. There are many other coatings available, depending on wall locations and other considerations, such as compatibility with surface materials, the amount of surface preparation involved, application methods and drying time, exposure of the finished surface to water, sunlight, temperature variations, chemicals, and potential physical abuse due to normal wear and tear. Figure 6.28 shows some of the possible paint and coating finishes and their applications.

Another common wall material in a commercial building is ceramic tile. (It is also used in flooring.) This material can be found in various applications throughout the building but is most common in wet areas such as restrooms and cafeterias. It is a durable material that withstands water exposure. This tile allows the designer many options in color and patterns

TYPE	PRINCIPAL BINDER	BASE/CURE	TYPICAL USES	COMPARATIVE COST RANGE	INSERVICE LIFE RANGE IN YEARS	GLOSS RETENTION	STAIN RESISTANCE	WEATHER RESISTANCE	ABRASION IMPACT RESISTRANCE	FLEXI-BILITY
Clear	Acrylic, methyl methacrylate copolymer	Solvent; water	Waterproofing and surface sealer against dirt retention, graffiti; for vertical surfaces of concrete, masonry, stucco; may be pigmented.	Moderate to high	5 to 10	Excellent to good	Fair	Excellent to good	Good	Good
	Alkyd, spar varnish	Solvent	For interior and protected exterior wood surfaces. Also as vehicle for aluminum pigmented coatings.	Moderate	Up to 1 exterior	Fair to good	Poor	Poor	Fair	Good
	Phenolic, spar varnish	Solvent	Exterior wood surfaces subject to moisture. May be used in marine environments. Also vehicle for aluminum pigment.	Moderate to high	Up to 2 exterior	Fair to good	Fair	Good	Good	Good
	Silicane	Solvent	Waterproofing and surface sealer against dirt retention for vertical surfaces of concrete, masonry, stucco.	Moderate	5 to 7	Flat	Fair	Good	Penetrating coating	
	Urethane, one-part	Moist cure1	Surfaces subject to chemical attack; abrasion, graffiti, heavy or concentrated traffic, such as gymnasium floors.	Moderate to high	up to 15	Excellent to good	Good to excellent	Good to excellent	Good to excellent	Excellent
Stain	Acrylic	Solvent; water	Pigmented translucent or semi-opaque exterior surface sealers; solvent based for masonry, concrete; water based for wood.	Moderate to low	3 to 5	Flat finish	Not a factor	Good to fair	Penetrating coatings – resistance for substrate	
	Alkyd	Solvent; water	Pigmented exterior or interior surface sealer for wood surfaces such as shingles, does not impart sheen to surface.	Moderate	3 to 6	Flat finish		Fair		
	Oil	Solvent	Pigmented exterior or interior or surface sealer or wood such as shingles, trim, opaque or semitransparent.	Moderate	3 to 5	Fair		Fair		
Opaque	Acrylic	Water	For exterior/interior vertical surfaces of wood, masonry, plaster, gypsum board, metals. Good color retention. Permeable to vapor.	Moderate to low	5 to 8	Good to fair	Fair	Good	Good to fair	Good to excellent
	Acrylic, epoxy modified, two-part	Water	High performance coating for interior vertical surfaces subject to graffiti, stains, heavy scrubbing. May be used in food preparation areas.	High	10 to 15	Good	Good	Good to excellent	Good to excellent	Good to excellent
	Alkyd	Solvent; water	For exterior/interior vertical and horizontal surfaces, such as wood, metals, masonry. Poor permeability to vapor.	Moderate	5 to 8	Good to excellent	Fair	Fair to good	Fair to good	Fair to good
	Chlorinated rubber	Solvent	Swimming pool coatings. Corrosion protection; isolating dissimilar metals.	High to very high	Up to 10	Fair	Fair	Good	Fair to good	Good
	Chlorosulfonated polyethylene	Solvent	Protective coating for tanks, piping, valves, elastomeric roofing membranes.	Very high	Up to 15	Not applicable	Fair	Excellent	Fair to good	Excellent
	Epoxy, two-part; epoxy ester, one part	Solvent cure; solvent	Moisture/alkali resistant. Two-part for nondecorative interior uses highly resistant to chemicals. Esters in wide choice of colors.	High to very high	15 to 20; Up to 10	Poor to good	Excellent for two-part	Good to excellent	Excellent	Good to excellent

Coating	Vehicle	Description						
Phenolic	Solvent	Chemical- and moisture-resistant coatings. May be used over alkaline surfaces	Moderate to high	Up to 10	Fair	Fair	Good to excellent	Good to excellent	Good
Polychloroprene	Solvent[2]	Marketed as "Neoprene"; resistant to chemicals, moisture, ultraviolet radiation. Also used as roofing membrane; generally covered with Hypalon	Very high	Up to 25	Not applicable	Good	Excellent	Excellent	Good
Polyester	Solvent	Limited application in field; over cementitious surfaces, metal, plywood for exterior exposures	High	Up to 15	Good to excellent	Good to excellent	Good to excellent	Good	Good to excellent
Silicone	Solvent	Surfaces with temperatures up to 1200 °F. Often with aluminum pigments. Corrosion and solvent resistant	Very high	Vanes	Not applicable, special purpose coating			Good	Good
Silicone; modified acrylic, alkyd, epoxy	Solvent	High-performance exterior coatings. Industrial siding, curtain walls, when shop-applied baked-on	High to very high	15 to 20	Good to excellent	Good	Good to excellent	Good to excellent	Good
Styrene, butadiene	Water	interior coating for gypsum board, plaster, masonry. Limited exterior use over cementitious substrate, as filler over rough porous surfaces	Moderate to low	4 to 8	Poor to fair	Fair	Poor	Fair	Good
Urethane, one or two part	Moist or chemical cure[3]	Heavy-duty wall and floor coatings. Resistance to stains, chemicals, graffiti, scrubbing, solvents, impact, abrasion	High to very high	15 to 20 -	Excellent	Good to excellent	Good to excellent	Good to excellent	Excellent
Vinyl, polyvinyl chloride-acetate	Solvent	Residential metal siding and trim, gutters, lead-ers, baseboard heating covers, when shop-applied, baked-on	High	Up to 15	Good	Fair	Good	Good	Good to excellent
Vinyl, polyvinylidiene chloride	Water	Metal and concrete surfaces in contact with dry and wet food, potable water, wastewater, jet and diesel fuels	High	Up to 10	Good	Fair	Good	Good	Good
Vinyl, polyvinyl acetate	Water	Exterior and interior vertical surfaces, such as masonry, concrete, wood, plaster, gypsum board, metals. Permeable to vapor.	Moderate to low	5 to 8	Good to fair	Fair	Good	Good to fair	Good
Bituminous, coal tar pitch, asphalt; emulsions, cutbacks	Solvent	Waterproofing of metals, concrete, masonry, portland cement plaster, piping when below grade or immersed	Low	10 to 15 Protected	Not a factor		Good	Poor	Fair
Cement	Water	Leveling coat over porous masonry or concrete not subject to abrasion or scrubbing. Cement and oil used as primers for metal surfaces	Low	varies	Flat finish	Poor	Poor for color	Good	Poor

[1] Solvent-based, oil-modified urethane is also available; for use on interior/exterior vertical and horizontal wood surfaces. Cost is moderate.
[2] May be obtained as water-reducible coating; use as field-applied coating very limited; generally used as tank linings.
[3] Solvent base, oil-modified urethane is also available; for use on vertical and horizontal surfaces. Cost is moderate, but durability is lower than for other types.

FIGURE 6.28 Paints and Coatings

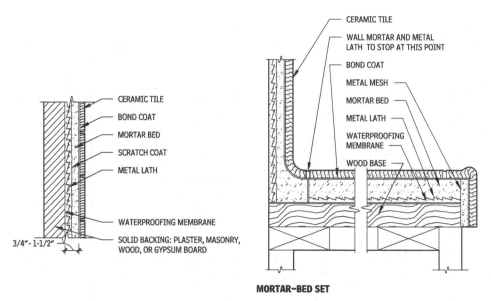

CERAMIC TILE
BOND COAT
MORTAR BED
SCRATCH COAT
METAL LATH

WATERPROOFING MEMBRANE
SOLID BACKING: PLASTER, MASONRY, WOOD, OR GYPSUM BOARD

3/4" - 1-1/2"

CERAMIC TILE
WALL MORTAR AND METAL LATH TO STOP AT THIS POINT
BOND COAT
METAL MESH
MORTAR BED
METAL LATH
WATERPROOFING MEMBRANE
WOOD BASE

MORTAR-BED SET

FIGURE 6.29 Ceramic Tile Applications

and can be applied over many base materials. Figure 6.29 shows several methods for applying this material to a commercial building. The details in this example will not typically be shown in the drawings. However, the patterns and colors applied to each wall and floor will be indicated in the drawings.

Floors

In addition to tile, other flooring finishing materials include resilient flooring such as vinyl-clad tile (VCT), carpeting, wood, and terrazzo.

Resilient flooring materials are easily maintained, are resistant to damage, and provide a quiet and comfortable floor surface due to their ability to give under foot pressure. The most common types of flooring materials are the vinyls. Figure 6.30 lists some of these materials and their qualities. Vinyl-clad tiles (VCT) come in the most variety of colors and textures and are very common in commercial buildings. Other vinyls are used as indicated by the requirements in the example. These materials are generally applied directly to the slab or base floor using some type of mastic. This will require proper preparation.

Wood floors are applied with basically the same techniques used in installing vinyls, as shown in Figure 6.31. These floors are typically used in low- to medium-traffic areas because they require a certain amount of maintenance. They do provide beautiful floors, especially when applied in patterns as indicated in the example. The downside is that they can be damaged easily and heavy traffic will deform the wood members over a period of time, necessitating their replacement. However, some commercial buildings will specify wood flooring for a certain aesthetic appearance.

Terrazzo is a type of flooring material that provides a durable, smooth floor that is easily maintained and can provide many varied aesthetic pos-

RESILIENT FLOORING CHARACTERISTICS

TYPE	BASIC COMPONENTS	SUB FLOOR APPLICATION		RECOMMENDED LOAD LIMIT (PSI)	DURABILITY	RESISTANCE TO HEEL DAMAGE	EASE OF MAINTENANCE	GREASE RESISTANCE	SURFACE ALKALI RESISTANCE	RESISTANCE TO STAINING	CIGARETTE BURN RESISTANCE	RESILIENCE	QUIETNESS
Homogeneous vinyl sheet	Vinyl resins and fillers	O	S	125-750	1	1	2-4	2	1	2-4	3-4	3	3
Solid vinyl tile	Vinyl resins and fillers	O	S	50-150	1-3	1	1-3	1-2	1	1-4	3	3	3
Vinyl composition tile	Vinyl resins and fillers	B, O	S	75	3	3	4	2	2	3	2	5	5
Rubber sheet without backing	Rubber compound, homogeneous or layered	O	S	50-75	1-3	1	3	5	3	3	1-2	2	2
Cork tile	Cork particles and resins		S	50-75	5	5	5	5	4	5	3	1	1
Rubber tile	Rubber compound, homogeneous or layered	-, O	S	50-75	1-3	1	3	5	3	3	1-3	2	2
Linoleum sheet and tile	Linseed oil, resins, cork dust, wood flour, filler on a backing, cork, wood floor, and oleoresins		S	75-250	3	4	3	2	5	3	1-3	3-4	2-3

FIGURE 6.30 Resilient Flooring

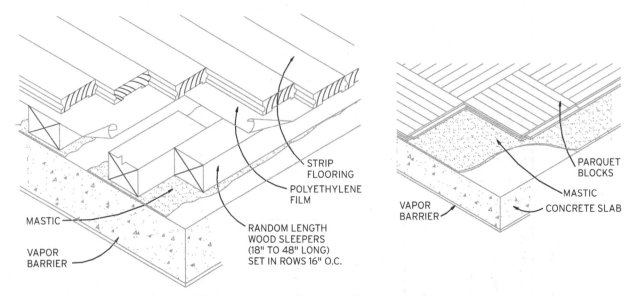

FIGURE 6.31 Wood Flooring Applications

sibilities. This material is a concrete compound with a topping of stone chips of a specific color and aggregate size to get the desired appearance. This material is more costly in materials and time to install, but it provides a much lower long-term maintenance expense than other flooring materials. The material is typically mixed and poured in place and leveled. Figure 6.32 shows some of the variations of this flooring material.

Carpets will be indicated the same as other flooring materials using the material codes and schedules, and will be applied to the building directly on the slab using a mastic, or on top of padding and attached to furring strips along the perimeter so the carpet can be stretched and fastened down. Once the flooring material is applied to the building, a cove base will be used to make the transition from floors to walls and to hide any irregularities between these two surfaces. Figure 6.33 shows some of the methods and materials used with various floor types to accomplish this transition.

Ceilings

Ceilings can consist of several types of systems. However, most ceilings in a typical commercial building will be either drywall applied as discussed earlier or, more commonly, as a suspended acoustical tile ceiling. This ceiling consists of a grid of main channels, cross tees, and splines suspended by wires from the structure overhead. Then acoustical tiles are placed into the grid for the finished ceiling and to hide the exposed structure and utilities above the room. In the space between the ceiling and roof deck, the plumbing piping, HVAC ductwork, and other utilities can be installed and remain invisible. The grid system and acoustical panels can be obtained in many colors and the tiles in various patterns. These ceiling grids usually are installed in either 2′×2′ or 2′×4′ patterns. These will be indicated in the drawings on the reflected ceiling plan, with the materials denoted in the specifications in Division 9. Figure 6.34 shows typical details of these

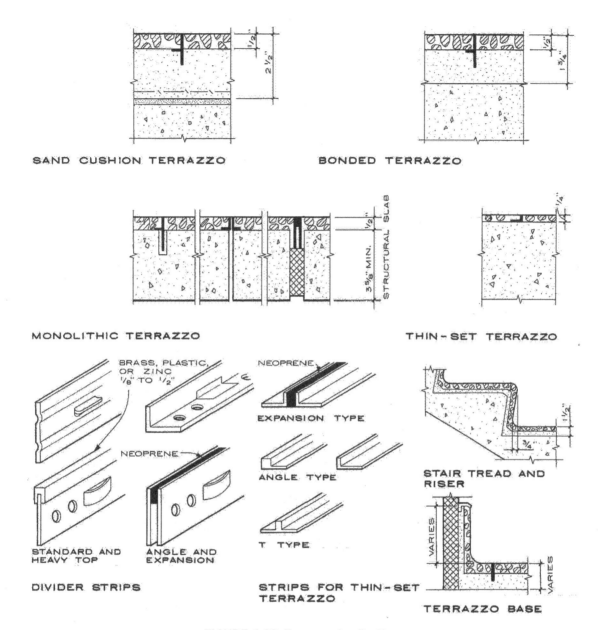

FIGURE 6.32 Terrazzo Applications

panels and how these systems are installed. Reflected ceiling plans are used to coordinate the lighting, HVAC supply and return grilles, fire protection heads, and other items applied to the ceiling to minimize conflicts during construction. Figure 6.35A shows a chart with ceiling grid material options and Figure 6.35B shows an actual legend that would be included on the reflected ceiling plan. The ceiling types can also be shown in the same manner as wall partitions to save space on the drawings.

DIVISION 10, "SPECIALTIES"

The definition of *specialty* items is those items that are typically purchased off the shelf in their final configuration, or only requiring minimal assembly, and are simply put in place in the building. The CSI MasterFormat

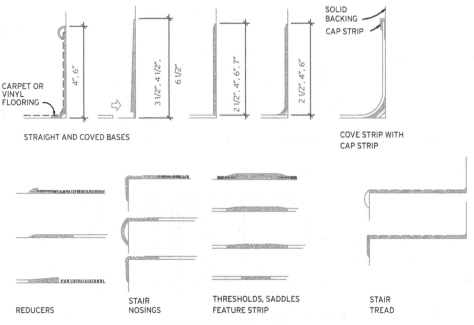

FIGURE 6.33 Cove Bases

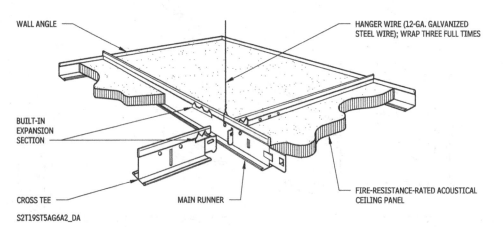

S2T19ST5AG6A2_DA

FIGURE 6.34 Lay-in Acoustical Ceilings

listing for Division 10 provides many of the items that are considered specialty items for today's commercial buildings. As you look at the list, you will note that most of these items can be purchased assembled and are installed or fastened into place in the building. Some require that certain construction in the building be made ready for their installation. Others can be bought and just fastened to the wall. In many cases, the owner will be able to negotiate the best price for these items and so will furnish them to the contractor to install. Very little may be included in the drawings regarding these items other that their general placement locations or if special mountings are required. Most of their definition will be in the specifications.

You will note in the CSI MasterFormat that a significant amount of specialty items are for restrooms, including the stall partitions, towel holders, grab bars, and so on. The most information contained in the drawings for

ACOUSTICAL CEILING SYSTEMS

CEILING TYPE	MAIN, CROSS T	ACCESS T'S	Z CHANNEL	H CHANNEL	T SPLINE	FLAT SPLINE	SPACER	MODULAR T	METAL PAN T	SPECIAL	BENT STEEL	BENT STEEL ALUM. CAP	BENT ALUMINUM	EXTRUDED ALUMINUM	GALVANIZED	PAINTED	ANODIZED	EMBOSSED PATTERN	FIRE RATING AVAILABLE	12 × 12	12 × 24	24 × 24	24 × 48	24 × 60	20 × 60	NOTES
	COMPONENTS										MATERIAL				FINISHES					ACOUSTIC TITLE SIZES (IN.)						
Gypsum wallboard	•										•				•	•			•							
Suspended	•																		•							
Exposed grid	•				•		•				•	•	•	•	•	•	•	•	•			•	•			
Semiconcealed grid	•				•	•					•				•	•			•			•	•			
Concelaed H & T				•							•				•	•			•							
Concelaed T & G			•								•				•	•			•	•						
Concealed Z			•			•					•				•	•			•	•						
Concealed access	•	•			•	•	•				•				•	•			•	•		•				
Modular	•				•			•			•				•	•			•					•	•	50 or 60" sq main grid
Metal pan									•				•				•			•	•					12" sq pattern
Linear metal										•			•			•										4" o.c. typical
Perforated metal	•						•				•		•			•										1 way grid 4'– 8' o.c.
Luminous ceilling										•			•			•	•									1" to 4" sq grid

FIGURE 6.35A Reflected Ceiling Plan Legends

123

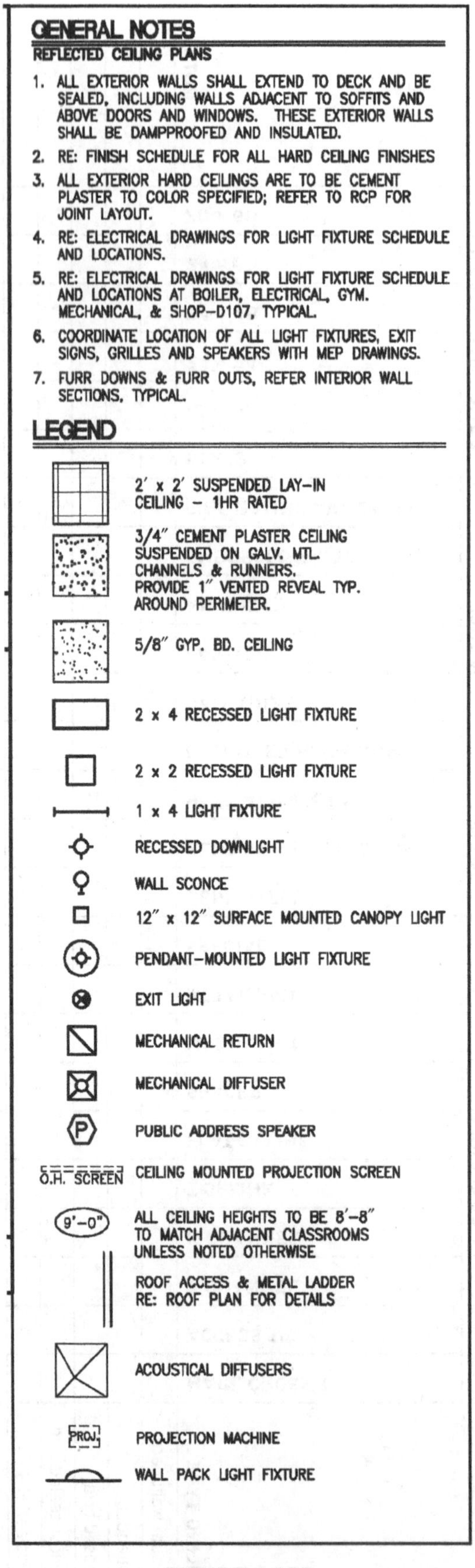

GENERAL NOTES
REFLECTED CEILING PLANS

1. ALL EXTERIOR WALLS SHALL EXTEND TO DECK AND BE SEALED, INCLUDING WALLS ADJACENT TO SOFFITS AND ABOVE DOORS AND WINDOWS. THESE EXTERIOR WALLS SHALL BE DAMPPROOFED AND INSULATED.

2. RE: FINISH SCHEDULE FOR ALL HARD CEILING FINISHES

3. ALL EXTERIOR HARD CEILINGS ARE TO BE CEMENT PLASTER TO COLOR SPECIFIED; REFER TO RCP FOR JOINT LAYOUT.

4. RE: ELECTRICAL DRAWINGS FOR LIGHT FIXTURE SCHEDULE AND LOCATIONS.

5. RE: ELECTRICAL DRAWINGS FOR LIGHT FIXTURE SCHEDULE AND LOCATIONS AT BOILER, ELECTRICAL, GYM. MECHANICAL, & SHOP—D107, TYPICAL.

6. COORDINATE LOCATION OF ALL LIGHT FIXTURES, EXIT SIGNS, GRILLES AND SPEAKERS WITH MEP DRAWINGS.

7. FURR DOWNS & FURR OUTS, REFER INTERIOR WALL SECTIONS, TYPICAL.

LEGEND

2′ x 2′ SUSPENDED LAY—IN CEILING – 1HR RATED

3/4″ CEMENT PLASTER CEILING SUSPENDED ON GALV. MTL. CHANNELS & RUNNERS. PROVIDE 1″ VENTED REVEAL TYP. AROUND PERIMETER.

5/8″ GYP. BD. CEILING

2 x 4 RECESSED LIGHT FIXTURE

2 x 2 RECESSED LIGHT FIXTURE

1 x 4 LIGHT FIXTURE

RECESSED DOWNLIGHT

WALL SCONCE

12″ x 12″ SURFACE MOUNTED CANOPY LIGHT

PENDANT—MOUNTED LIGHT FIXTURE

EXIT LIGHT

MECHANICAL RETURN

MECHANICAL DIFFUSER

PUBLIC ADDRESS SPEAKER

O.H. SCREEN   CEILING MOUNTED PROJECTION SCREEN

9′-0″   ALL CEILING HEIGHTS TO BE 8′-8″ TO MATCH ADJACENT CLASSROOMS UNLESS NOTED OTHERWISE

ROOF ACCESS & METAL LADDER RE: ROOF PLAN FOR DETAILS

ACOUSTICAL DIFFUSERS

PROJ.   PROJECTION MACHINE

WALL PACK LIGHT FIXTURE

FIGURE 6.35B

specialty items will also pertain to the restroom items and will be shown as indicated in Figure 6.36, where mounting heights are shown. The restroom plan views would provide the locational dimensions from walls and fixtures; a combination of information from both drawings will show the contractor all the installation dimensions.

Another common specialty item shown in the drawings is lockers. An interior elevation of the lockers installation and the location of the locker bank on a plan view would typically be all the information provided in the drawings.

DIVISIONS 11 AND 12, "EQUIPMENT" AND "FURNISHINGS"

Equipment and furnishings are typically not detailed to any great extent in the drawings, other than to show locations in order to aid the contractor in understanding why certain construction details are needed in specific locations.

To distinguish equipment from furnishings, we need to understand their use. Equipment can be defined as those items needed to conduct the business that the building is being constructed for. Typical plan views for equipment rooms include kitchen areas, laboratories, and other areas where equipment is placed to allow the building to function. These plan views would also be used in the mechanical and electrical categories of the working drawings for the consulting engineers to use in designing those building systems. Lockers would be equipment.

Furnishings are considered those items that are in place to accommodate the occupants, such as seating, modular furniture for office space, and items of decoration such as paintings, carpets, and even interior plants and vegetation.

Equipment could be furnished by either the owner or the contractor but is usually installed by the contractor. Furnishings will be typically purchased and even installed by the owner and, unless there are specific installation requirements for equipment, most of these items are shown on the plan views only for reference.

DIVISION 14, "CONVEYING EQUIPMENT"

Due to the current requirements for accommodating those with disabilities, most modern buildings have some method of conveying people, as well as goods, from one level to another. There are many types of conveying equipment available to accomplish this.

Moving goods can be accomplished with conveyors, lifts, dumb waiters, and hydraulic elevators. People are generally moved using elevators, escalators, or moving walks.

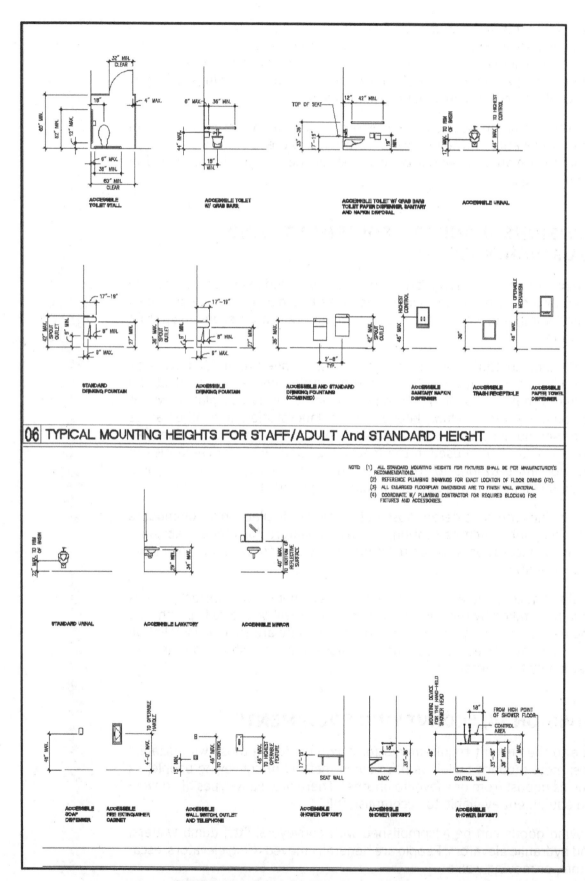

FIGURE 6.36 Accessible Mounting Heights

There are two typical types of elevators: electric and hydraulic. Elevators are more often used because they require little space and can move over a greater vertical distance that other conveyances. Hydraulic elevators can also be used for freight movers, since they can carry more weight than electric elevators. However, they are typically slower and are only able to cover a few levels or floors due to the cost of the cylinder used to move the cab. Electric elevators move much quicker and can cover many more levels, but they are limited in the amount of weight they are allowed to carry. Figure 6.37 and Figure 6.38 show illustrations of both electric and hydraulic types of elevators.

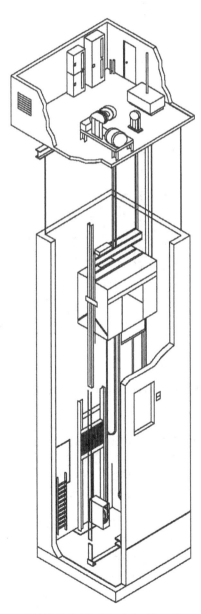

FIGURE 6.37 Electric Elevator

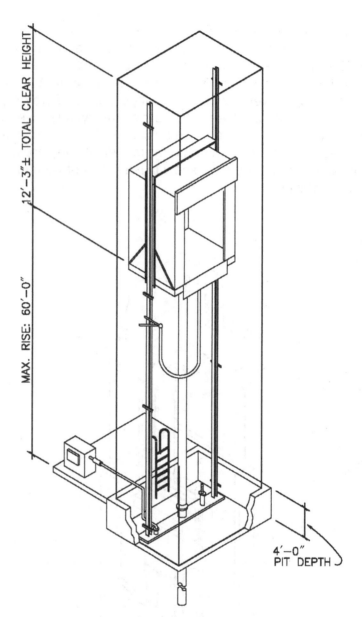

FIGURE 6.38 Hydraulic Elevator

Escalators can be used where there is enough horizontal room and no need for significant vertical travel. The primary application for these devices is in retail construction, where people are moved at a slower rate and the surrounding building is visible for viewing.

These conveyances are shown on the plan views and with some section views to show how they are integrated into the building construction. They are also shown in the mechanical and electrical categories because those support systems to allow them to operate. They will also be noted in the structural category since they will typically require special structural members for support.

SUMMARY

Now that we have learned something about the type of information that we typically find in the architectural category of the working drawings, we need to practice finding these pieces of information using the techniques we studied in Section I. Always begin the search by asking: What information am I looking for? Is it related to size, shape, and location, or materials, methods, and quality? This will allow you to select the document most likely to contain the information to begin your search. Then, using the structure of the selected document, work your way through the information to find your answer.

7

STRUCTURAL CATEGORY OF THE DRAWINGS

OVERVIEW

The structure of a commercial building is the skeleton of materials that all of the rest of the building is set or hung upon. On most commercial construction projects, there are some basic materials and practices used to construct this skeleton and some established practices for showing these building elements in the drawings.

The structural design begins with the geotechnical information, or Soils Report, as shown in Figure 7.1. This report is completed by a drilling/boring contractor and an interpretation provided by a Soils Engineer that describes the soil conditions and bedrock depth across the construction site. This report will assist the structural engineer in designing the structural members and components necessary to support the building and everything that will go into the building. This especially pertains to the structural elements that are at grade or below. The structural engineer will use the soils report when designing piers and grade beams and determining whether the soils will provide the support needed to balance the weight of the building and its contents.

Before we start into the specifics of the structure, let's discuss the basic structural components for a modern commercial building by examining Figure 7.3. In most cases, the building structure will begin with some type of pier that is drilled into the earth to bedrock. Next, grade beams are placed to span the distance from pier to pier. At this point, a slab can be poured and columns erected to start the first floor. At the next level of

RELATED SPECIFICATION DIVISIONS

Division 3, "Concrete"

Division 4, "Masonry"

Division 5, "Metals"

Division 6, "Woods, Plastics, and Composites"

Print and Specifications Reading for Construction, Updated Edition. Ron Russell.
© 2024 John Wiley & Sons, Inc. Published 2024 by John Wiley & Sons, Inc.
Companion website: www.wiley.com/go/printspecreadingupdatededition

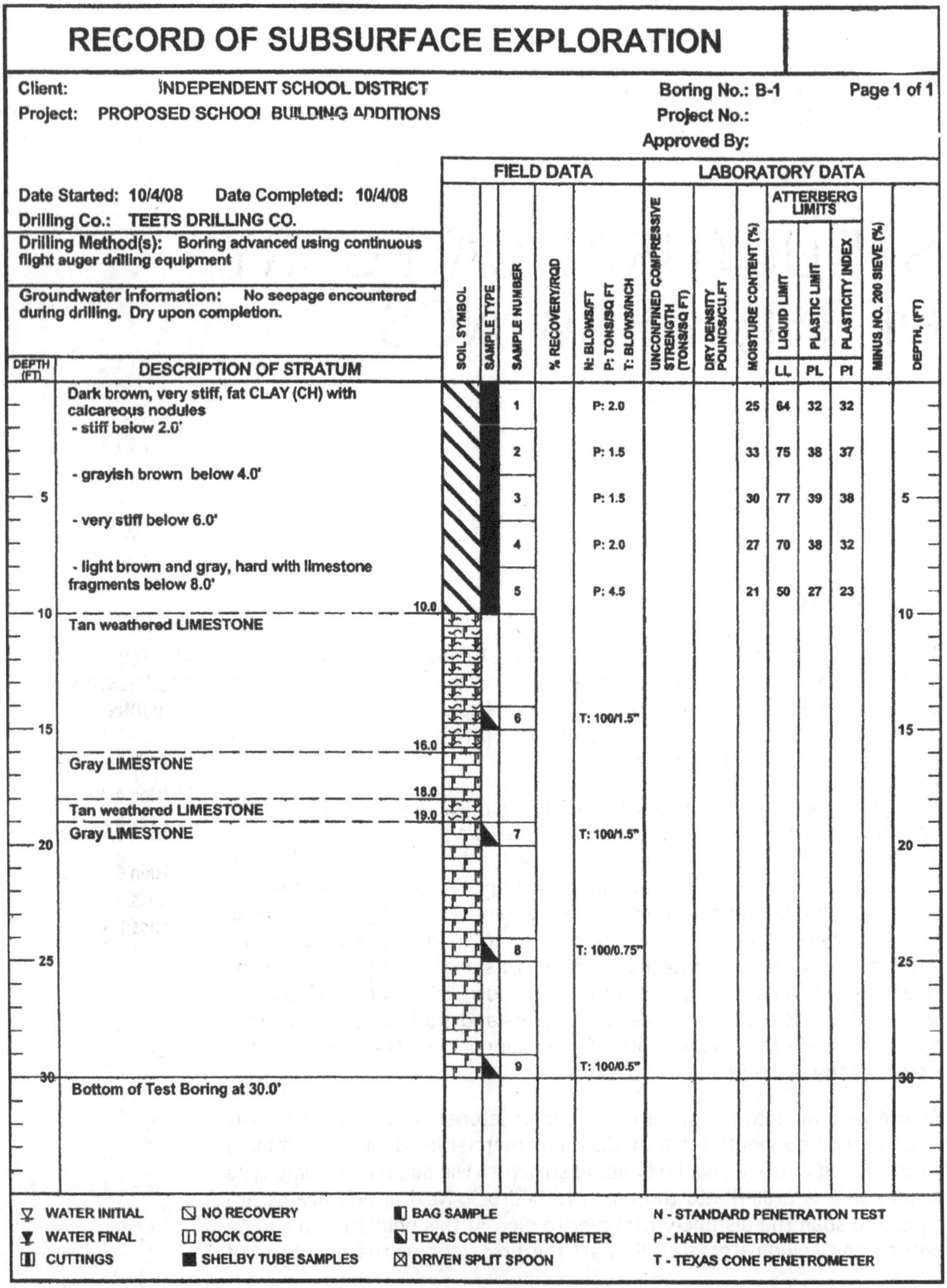

RECORD OF SUBSURFACE EXPLORATION

Client: INDEPENDENT SCHOOL DISTRICT

Project: PROPOSED SCHOOL BUILDING ADDITIONS

Boring No.: B-1   Page 1 of 1

Project No.:

Approved By:

Date Started: 10/4/08   Date Completed: 10/4/08

Drilling Co.: TEETS DRILLING CO.

Drilling Method(s): Boring advanced using continuous flight auger drilling equipment

Groundwater Information: No seepage encountered during drilling. Dry upon completion.

						FIELD DATA			LABORATORY DATA				ATTERBERG LIMITS				
DEPTH (FT)	DESCRIPTION OF STRATUM	SOIL SYMBOL	SAMPLE TYPE	SAMPLE NUMBER	% RECOVERY/RQD	N: BLOWS/FT P: TONS/SQ FT T: BLOWS/INCH	UNCONFINED COMPRESSIVE STRENGTH (TONS/SQ FT)	DRY DENSITY POUNDS/CU.FT	MOISTURE CONTENT (%)	LIQUID LIMIT LL	PLASTIC LIMIT PL	PLASTICITY INDEX PI	MINUS NO. 200 SIEVE (%)	DEPTH (FT)			
	Dark brown, very stiff, fat CLAY (CH) with calcareous nodules - stiff below 2.0'			1		P: 2.0			25	64	32	32					
				2		P: 1.5			33	75	38	37					
5	- grayish brown below 4.0'			3		P: 1.5			30	77	39	38		5			
	- very stiff below 6.0'			4		P: 2.0			27	70	38	32					
	- light brown and gray, hard with limestone fragments below 8.0'			5		P: 4.5			21	50	27	23					
10	10.0 Tan weathered LIMESTONE													10			
15				6		T: 100/1.5"								15			
	16.0 Gray LIMESTONE																
	18.0 Tan weathered LIMESTONE 19.0																
20	Gray LIMESTONE			7		T: 100/1.5"								20			
25				8		T: 100/0.75"								25			
30				9		T: 100/0.5"								30			
	Bottom of Test Boring at 30.0'																

▽ WATER INITIAL	◩ NO RECOVERY	▯ BAG SAMPLE	N - STANDARD PENETRATION TEST	
▼ WATER FINAL	▯ ROCK CORE	◣ TEXAS CONE PENETROMETER	P - HAND PENETROMETER	
▯ CUTTINGS	■ SHELBY TUBE SAMPLES	⊠ DRIVEN SPLIT SPOON	T - TEXAS CONE PENETROMETER	

FIGURE 7.1 Geotechnical Information

structural members, girders, or beams will be attached to the columns to connect the structural elements. Open web joists are then used to span the distances from beam to beam and to close up the space between the columns so decking can be applied. Once past the piers and grade beam, this process of installing columns and beams for each subsequent floor can be continued as needed and designed by the structural engineer. The intent of the structural design is that weight loads are transferred from member to member, with these loads accumulating until the total weight is supported by the bearing substrate.

When looking at the specific structural members used on a building, it is recommended that the reader of the plans begin by examining the first sheets in the structural category of drawings, S1 and/or S2. Typically, the structural engineer will use these first sheets to provide general notes that define the design loads he calculated for to provide the definitions of materials and performance criteria for each structural member. The reader should always read these general notes before proceeding through the structural drawings. In Figure 7.2 we can see some of the typical symbols used to depict some of the structural elements used in a set of plans. Also shown are the typical symbols used to depict the various structural materials used in section views in the plans. We will see other examples of nomenclature and symbols used as we discuss each structural member.

Let's look at each of these components, the materials used to construct them, and how they are represented on the drawings.

DIVISION 3, "CONCRETE"

Most structures start with some type of foundation or footings attached to the earth, and the foundation is usually made of concrete. Each of these components will have to carry tremendous loads of the building to function.

Piers

As mentioned before, the structure begins with piers that are cored down into the supporting substrate, which is usually some type of bedrock as indicated by a geotechnical report like the one shown in Figure 7.1. Most piers are constructed by drilling the earth to a depth required to reach the bedrock, inserting reinforcing steel into the drilled hole, and filling the hole with concrete. The piers are designed by the structural engineer to support a specific load from the building, plus live loads from the elements and any load-associated facility activities that the owner will conduct within the building. As you can see in Figure 7.3, the weight of the entire structure is dependent on the piers.

At the time of construction, the actual depth to the bearing substrate may not be known. The owner will provide an allowance for the contractor to use for each foot of depth he must core until he reaches the bedrock that will support the building. As a result, the structural engineer will develop

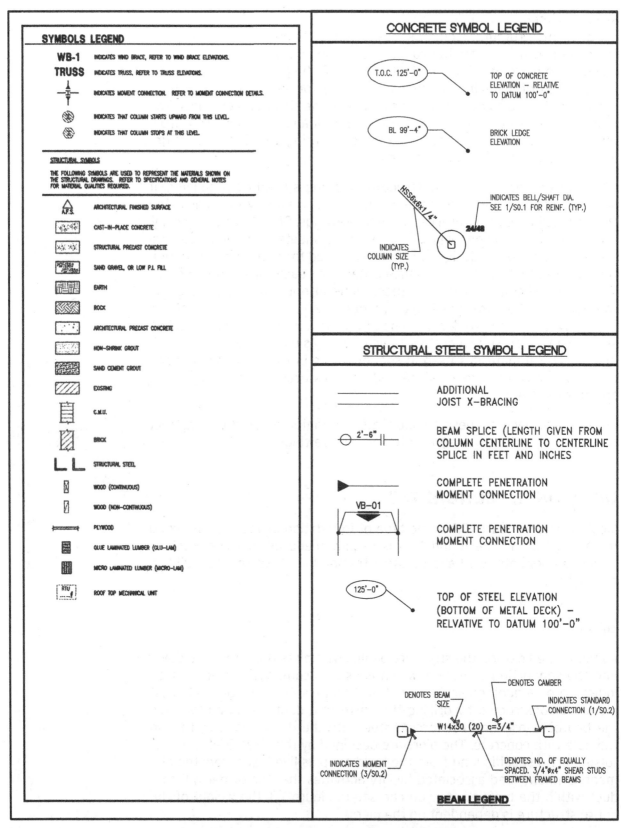

FIGURE 7.2 Symbols Used for Structural Materials

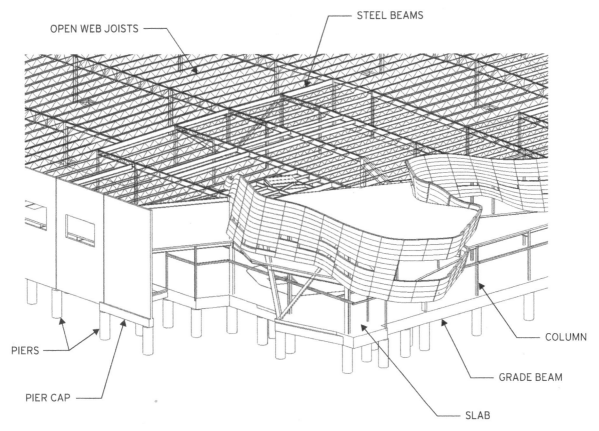

OPEN WEB JOISTS

STEEL BEAMS

PIERS

PIER CAP

COLUMN

GRADE BEAM

SLAB

FIGURE 7.3 Structural Elements of a Building

a typical pier design that will allow the contractor to bid the project. Figures 7.4 and 7.5 show how these details are typically shown in the plans.

Figure 7.4 shows a typical belled pier with its diameter and reinforcing requirements shown in the schedule. This typical pier shape can be used for each pier. Those reading the plans would need to know which pier they are looking at by identifying it from the plan view using its callout as a P1, P2, or P3 pier. Knowing the callout allows readers to look at the pier schedule and determine that, if it were a P1 pier, for example, the shaft is 18 inches in diameter, the bell is 54 inches in diameter, and it requires six number 5 vertical reinforcing bars. Reinforcing bars are numbered by their nominal diameter. For example, for P1 piers number 5 bars are required; that means the bars are $5 \times 1/8''$ in diameter, or $5/8''$ in diameter. Figure 7.6 shows us more of the identifying nomenclature for reinforcing bar.

Figure 7.5 shows another typical pier detail that is used for bidding purposes. To know exactly how deep a pier will be, the contractor must refer to the plan views in the structural category and locate the pier using the identification marks shown, determine the top elevation of the pier, and compare that to the depth to the bearing substrate, using the geotechnical reports if available. The rest of the information for each pier will be provided in the schedule.

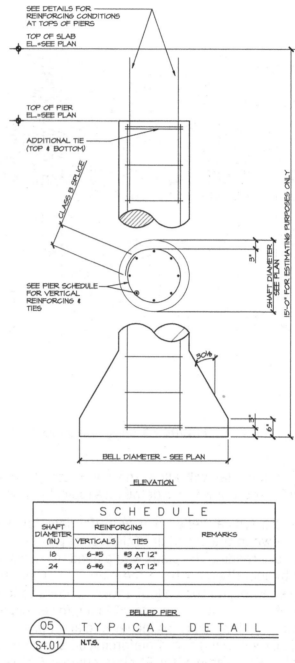

SEE DETAILS FOR REINFORCING CONDITIONS AT TOPS OF PIERS

TOP OF SLAB EL.=SEE PLAN

TOP OF PIER EL.=SEE PLAN

ADDITIONAL TIE (TOP & BOTTOM)

CLASS B SPLICE

SEE PIER SCHEDULE FOR VERTICAL REINFORCING & TIES

SHAFT DIAMETER SEE PLAN

15'-0" FOR ESTIMATING PURPOSES ONLY

3"

30°

3"

6"

BELL DIAMETER - SEE PLAN

ELEVATION

SCHEDULE			
SHAFT DIAMETER (IN.)	REINFORCING		REMARKS
	VERTICALS	TIES	
18	6-#5	#3 AT 12"	
24	6-#6	#3 AT 12"	

BELLED PIER

05 S4.01 — TYPICAL DETAIL — N.T.S.

FIGURE 7.4 Bell Pier Detail

Grade Beams

Once the piers are completed, they are connected using grade beams. These structural members are called *grade beams* because they are an actual beam that is located at grade, or at the level of the soils. These beams are typically rectangular in shape and span the distance from pier to pier. Figure 7.7 shows a typical section view of a grade beam. Notice the void box indicated at the bottom of the beam used to protect it from the soils underneath that could swell with moisture. These beams take on many different shapes and configurations to facilitate the use of various construction materials and to allow creation of different-shaped buildings. We will see examples of this as we move forward in this unit.

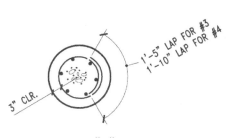

SECTION "A"

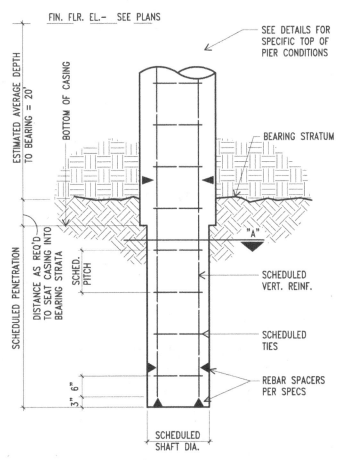

TYPICAL PIER DETAIL CASED PIER
NO SCALE

DRILLED PIER SCHEDULE						
MARK	PIER DIAMETER (IN.)	PENETRATION (FT.)	VERTICAL REINFORCING	TIES	CAPACITY	NOTES
P1	18	4	6#5	#3@12		

FIGURE 7.5 Shaft Pier Detail

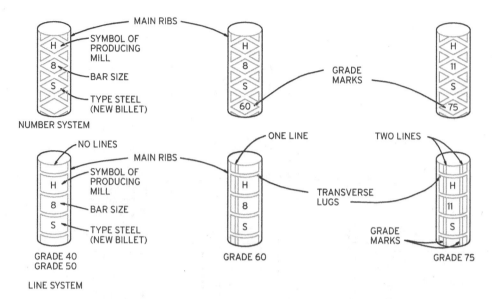

FIGURE 7.6 Reinforcing Bar Grade Mark Identification

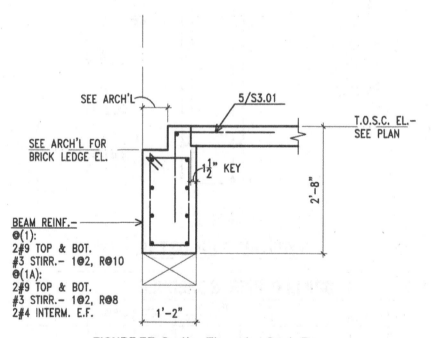

FIGURE 7.7 Section Through a Grade Beam

Slab

There are several types of slabs used in commercial construction, the slab-on-grade and structural slabs being most common. The slab-on-grade is used more often and is the simplest to define in the plans. In fact, most of the slab definition is provided in the general notes on the front sheets of the structural category of drawings. However, if you look at Figure 7.7 again, you can see the slab shown in reference to the grade beam. The slab is typically shown this way in the drawings because its shape does not require a lot of detailing.

Once the piers, grade beams, and slab are in place, we can begin looking at the structural members that are above the grade. At this point, depending on the requirements of the owner and the building, the structural engineer can take advantage of several different types of materials to complete the structure. He could, for example, continue utilizing concrete as the material of choice and construct columns, girders, and beams by pouring them in place. However, the engineer can also utilize metals, masonry units, and wood to continue the structure. We will look at the use of metals first since this is the primary material used for these components on most commercial projects.

DIVISION 4, "MASONRY"

If we look at Figure 7.3 again, we can see that one of the sides of the structure is noted as being masonry or wood. By understanding how some of the connections are made, we can transition from one material to another to accomplish unique building designs and still have a structure that performs properly. As we noted in the previous paragraph, we can transition from a metal structure to a masonry structure by using embedded metal plates that allow us to secure beams, girders, and joists to the masonry wall. This wall is then constructed to transfer weight as a unit.

Using masonry units for structural walls allows the designer to accomplish several things at once. The designer can structurally support the building design and, using the appropriate bond pattern for the units, can achieve a patterned finish for the walls. Figure 7.8 shows how some of these bond patterns can be achieved using brick as the masonry unit and interlocking two wythes together to obtain a structural wall. However, brick today is primarily used as a finish; most modern commercial buildings utilize concrete block as the masonry unit of choice. We will look at the techniques for that type of construction.

Masonry units allow the architect some creativity in utilizing different materials to create different textures and appearances while still creating a sound structure. These units can also be insulated for thermal and sound transfer, sealed to provide a waterproof barrier, and painted, the same as with any wall. The only thing that must be accomplished is that the wall must be built correctly to provide structural qualities.

Figure 7.9 shows some of the requirements for this type of construction. Unlike the metal structure where the weight is transferred from member to member, the weight in a masonry wall will travel in a straight line until it reaches an opening, or a beam or footing. At the openings is where the structural engineer will have to specify how to use the units to facilitate the weight transfer, as shown in the example. Notice that the course directly above the opening for the door is filled with grout or concrete and reinforcing to create a solid unit. This is called a bond beam lintel. Then the open cores of the masonry units on each side of the opening are filled

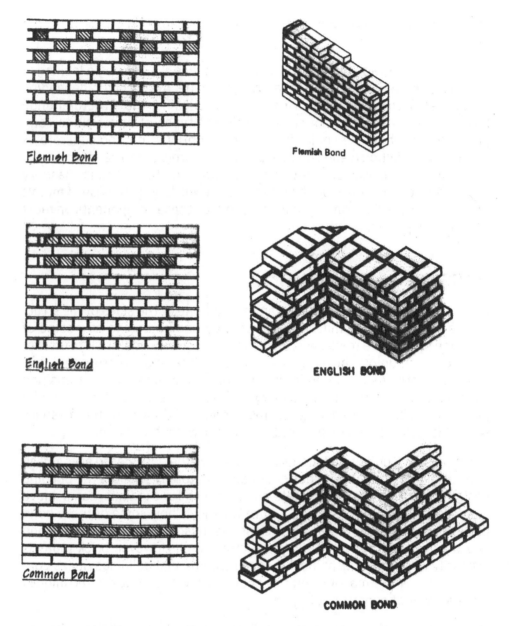

FIGURE 7.8 Masonry Bonds Typically Used in Structural Masonry

and reinforced to create solid bearing units that will allow the weight the wall is carrying to transfer around the opening. These solid bearing units function very much like the metal beams and girders. At the top of the masonry wall and all along its perimeter, the top course is filled with reinforcing and concrete to create a bond beam. This bond beam acts as a top structural beam or girder and is where the metal plates would be embedded to allow the metal beam and joists to be attached to the masonry units, thus transferring from one structural material system to another.

These beams can be created at various heights throughout the elevation of the wall to create structural strength and the cores filled vertically at various distances along the length of the wall to create the affect of columns.

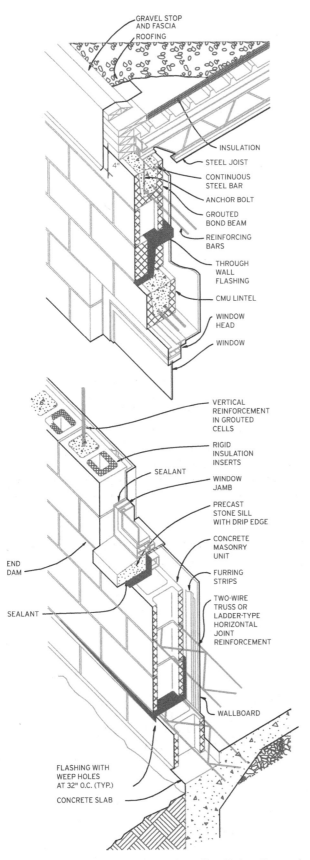

GRAVEL STOP
AND FASCIA

ROOFING

INSULATION

STEEL JOIST

CONTINUOUS
STEEL BAR

ANCHOR BOLT

GROUTED
BOND BEAM

REINFORCING
BARS

THROUGH
WALL
FLASHING

CMU LINTEL

WINDOW
HEAD

WINDOW

VERTICAL
REINFORCEMENT
IN GROUTED
CELLS

RIGID
INSULATION
INSERTS

SEALANT

WINDOW
JAMB

PRECAST
STONE SILL
WITH DRIP EDGE

CONCRETE
MASONRY
UNIT

FURRING
STRIPS

TWO-WIRE
TRUSS OR
LADDER-TYPE
HORIZONTAL
JOINT
REINFORCEMENT

END
DAM

SEALANT

WALLBOARD

FLASHING WITH
WEEP HOLES
AT 32" O.C. (TYP.)

CONCRETE SLAB

FIGURE 7.9 Masonry Structural Construction Using Concrete Block

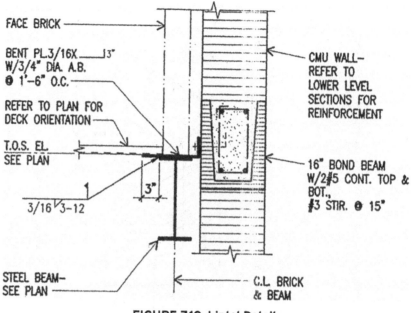

FIGURE 7.10 Lintel Detail

The drawings will typically show the masonry walls in plan views using the symbol shown in Figure 7.2. Then, typical details will be shown for the mason to use in creating the lintels and bond beams, such as in Figures 7.10 and 7.11. The specifications will tell the mason how often he must create the bond beams and fill cell columns for this structure. It is then up to the mason to locate these beams and lintels as needed and to build them as defined in the drawings. Note in Figure 7.11 the weld symbol used to show how the joist is to be secured to the embedded plate of the bond beam.

Much of the theory of weight transfer in masonry construction is true in wood framing construction as well. Division 6 discusses using wood in structures.

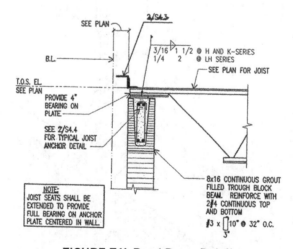

FIGURE 7.11 Bond Beam Detail

DIVISION 5, "METALS"

Before we talk about how the metal components are used in the building structure, we need to understand the various shapes available and how symbols and nomenclature are used in the drawings to call them out.

Look at Figure 7.12. This example shows the various shapes of metal beam materials, angles, and tubing, which come in various sizes that the structural engineer can use in design. The best way to understand the nomenclature is to look at the wide-flange shape. In the example, the beam material is called out as a W18×77. The W means that the beam is a W section metal member in reference to its specific shape, that it is nominally 18 inches tall (or deep), and that it weighs 77 pounds per linear foot. This weight per linear foot is how the structural engineer determines which sizes of beams to use in each application: The more weight, or metal per foot, the more the beam is capable of supporting. Square tubing and pipe are specified by their nominal outside dimensions and the thickness of their walls. Plate is specified by its thickness. Understanding this will allow us to identify the various members on the drawings. Let's first look at columns as we continue building the structure.

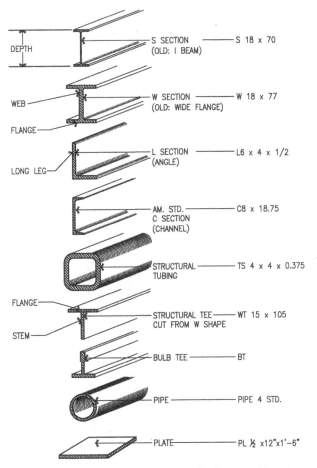

FIGURE 7.12 Steel Shapes Commonly Used In Structures

Columns

Columns can be fabricated from W or S sections, angles, channels, or tubing and pipe, similar to the examples shown in Figure 7.13. In Figure 7.13, a base plate and top plate are welded to each end of the members being used so that the column load is distributed to foundation members and the total column is at the correct height for the location it will be placed at in the structure. In Figure 7.14 we can see typical column base plate details

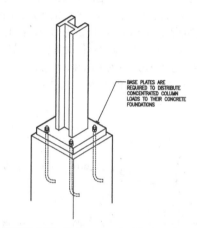

BASE PLATES ARE REQUIRED TO DISTRIBUTE CONCENTRATED COLUMN LOADS TO THEIR CONCRETE FOUNDATIONS

FIGURE 7.13 Columns and Base Plates

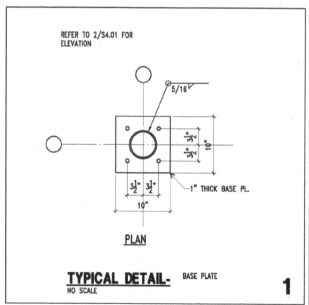

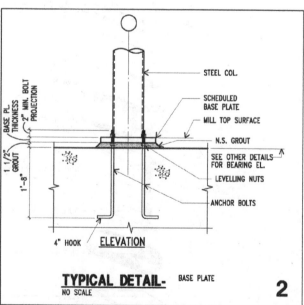

FIGURE 7.14 Column Details Used In Drawings

COLUMN SCHEDULE

MARK	SIZE	BASE PLATE, U.N.O. ON PLAN
C1	5"⌀ STD PIPE	BP1

BASE PLATE SCHEDULE

MARK	LENGTH	WIDTH	X	Y	THICKNESS	ANCHOR BOLTS	NOTES
BP1	11	11	3 1/2	3 1/2	1	4 – 3/4"⌀ X 2'–0"	1/S4.01

4

5

FIGURE 7.15 Column and Base Plate Schedule

that we would find in most drawing sets. It shows how the plates are applied to a square tube or round pipe by welding it to these shapes. Anchor bolt-holes are located by dimensions that are signified by letters corresponding to each specific column in the schedule shown in Figure 7.15. If we were using a column C1 at a certain location, we could look at the schedule and determine what our dimensions should be for the overall size of the plate and the location of the boltholes. The schedule also indicates the size of anchor bolts to be used for each column. The plate will be welded all around on all sides to the metal column.

Welding on a commercial construction project must be accomplished by a certified welder. He will read the symbols shown in the plans to determine which welds must be used on each member. Let's study Figure 7.16(A), where welding symbols are explained in detail. The structural engineer will determine the type of weld required for each connection. Figure 7.16(B) shows some of the typical welds used for metal structural members for a commercial building. The welder looking at Figure 7.14 would see that a 5/16″ fillet weld is required all around the tubing or pipe to connect the base plate to the column.

Once the column is completed, it can be assembled in place as shown in Figure 7.17. Anchor bolts will have been embedded into the beam or pier where the column is to be located, leveling nuts placed on the bolts to level the column, top bolts placed on top of the pate to lock the column in place, and then an epoxy grout used to secure the leveling nuts in place. This is also shown in Figure 7.14 on the typical base plate detail. Figure 7.17 shows how this assembly fits in place on the pier or beam to continue the structure, which has now undergone a transition from concrete to metal elements.

BASIC WELD DEVICE SYMBOLS

BACK WELD	FILLET WELD	PLUG OR SLOT	GROOVE	OR BUTT JOINTS					
			SQUARE	V	BEVEL	U	J	FLARE V	FLARE BEVEL
⌒	◺	▭	‖	V	V	Y	Y	V	⫦

SUPPLEMENTARY WELD SYMBOLS

BACKING	SPACER	WELD-ALL-AROUND	FIELD WELD	FLUSH	CONVEX
[M]	—[M]—	◯	⚑	—	⌒

NOTE: For additional basic and supplementary weld symbols, see the American Welding Society A2.4-79.

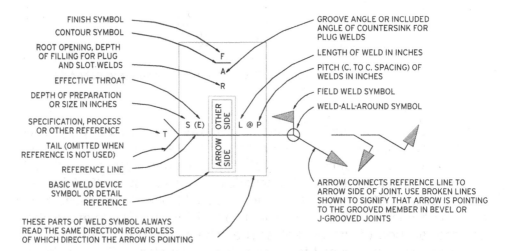

FIGURE 7.16(A) Welding Symbols

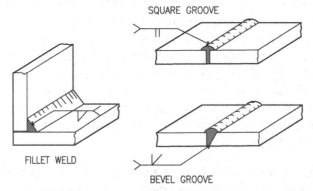

FIGURE 7.16(B) Welds Used on Construction Projects

Beams and Girders

With our columns in place, we can now start connecting the rest of the structure to them, typically using beams and girders. Girders are a type of beam that is usually larger in size and designed to span greater lengths than beams, both of which are typically W and S section members. As shown in Figure 7.12, these members can come in many different sizes and

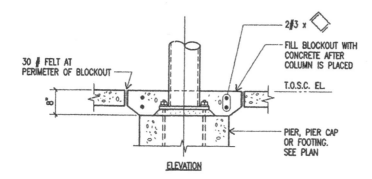

TYPICAL DETAIL
NO SCALE

FIGURE 7.17 Column in Place on Piers

lengths. These members are fabricated at the mill and then assembled in the field at the locations shown on the drawings. Lengths and sizes are determined by the structural engineer to support the design loads. The connections used to assemble each beam into place vary, depending on the location and sizes of adjoining beams.

Figure 7.18 shows some typical methods for connecting beams and girders to the tops of columns and to each other. The types of connections shown indicate the types of structural steel framing defined by The American Institute of Steel Construction. Rigid connections are based on the beam-column connections being able to hold their original angles under load. Shear connections are based on the beams and girders being connected for shear only and can move freely under twisting loads. Semirigid beam and girder connections possess a small but known moment-resisting capacity.

These connections are typical for the types used on structural steel members on a commercial construction project. However, there are many different configurations that can be used by the structural engineer to solve the problems created by modern architectural designs and still achieve the load support defined by the three types. The connection types that will be generally used for the project will be defined by the structural engineer, but each specific connection will be detailed on the shop drawings.

Open Web Joists

To close in the space created between the beams and girders, open web joists are used. These joists are constructed from the various different metal shapes shown in Figure 7.12 to create the desired structural

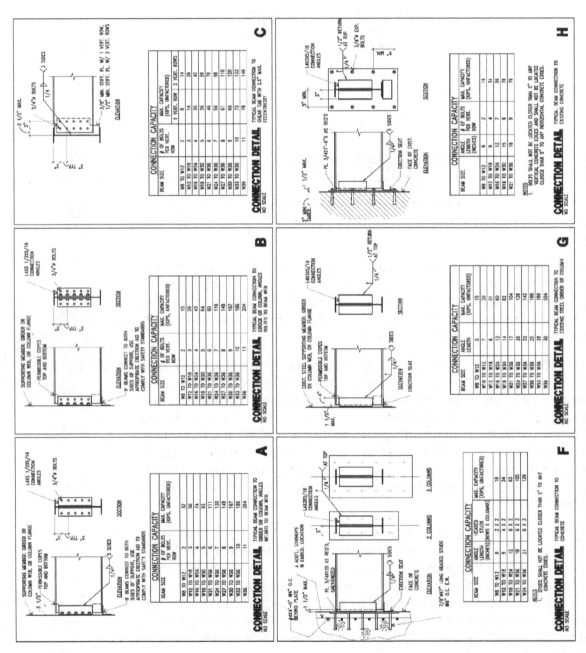

FIGURE 7.18 Methods for Connections for Beams, Girders, and Columns

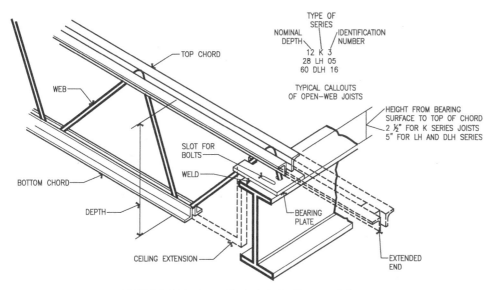

FIGURE 7.19 Open Web Joists Nomenclature

member. Figure 7.19 shows us the nomenclature for these structural elements.

Open web joists come in three typical series: K standard, LH longspan, and DLH deep longspan. These series define depth of the joists and the maximum spans, as shown in the schedule in Figure 7.20. These designations also define the bearing plate requirements. Figure 7.20 shows the varying heights required for each series and the bearing plate lengths needed when used with the different types of structural materials.

These joists are used between the beams and girders, or wood and masonry walls, as shown in Figure 7.21. They allow us to provide support across these spans for flooring or roof decking. Their spacing is determined by the structural engineer based on loads, which dictates how many of a specific type are required from beam to beam.

The joists not only provide support for decking materials but also allow for piping, electrical conduits, and HVAC ducting to be supported from or even passed through for service runs. Figure 7.22 shows how this is typically achieved in most commercial buildings. The bearing plates will be welded to supporting beams and girders, similar to what is shown in Figure 7.23. On masonry walls, metal plates will be imbedded in the masonry construction so the bearing plates can be welded to them.

Let's now look at that type of structural construction.

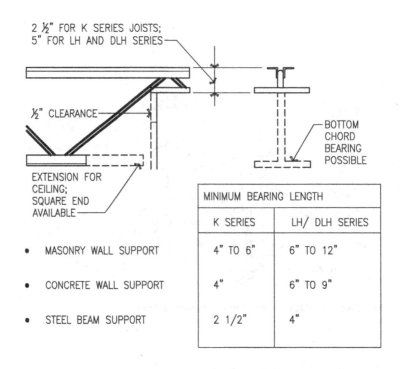

2 ½" FOR K SERIES JOISTS;
5" FOR LH AND DLH SERIES

½" CLEARANCE

BOTTOM CHORD BEARING POSSIBLE

EXTENSION FOR CEILING; SQUARE END AVAILABLE

- MASONRY WALL SUPPORT
- CONCRETE WALL SUPPORT
- STEEL BEAM SUPPORT

MINIMUM BEARING LENGTH		
	K SERIES	LH/ DLH SERIES
MASONRY WALL SUPPORT	4" TO 6"	6" TO 12"
CONCRETE WALL SUPPORT	4"	6" TO 9"
STEEL BEAM SUPPORT	2 1/2"	4"

OPEN WEB STEEL JOISTS		
STANDARD	K SERIES	8" TO 30" DEPTHS SPANNING UP TO 60'–0"
LONGSPAN	LH SERIES	18" TO 48" DEPTHS SPANNING UP TO 96'–0"
DEEP LONGSPAN	DLH SERIES	52" TO 72" DEPTHS SPAINNING UP TO 144'–0"

FIGURE 7.20 Open Web Joists Size Schedules

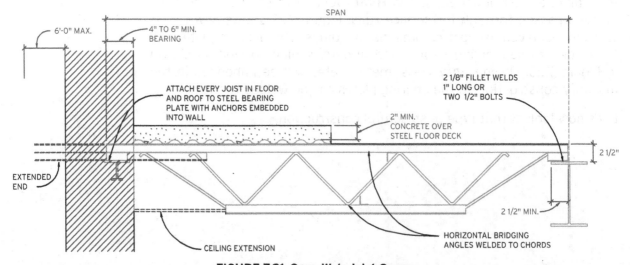

SPAN

6'-0" MAX.

4" TO 6" MIN. BEARING

ATTACH EVERY JOIST IN FLOOR AND ROOF TO STEEL BEARING PLATE WITH ANCHORS EMBEDDED INTO WALL

2 1/8" FILLET WELDS 1" LONG OR TWO 1/2" BOLTS

2" MIN. CONCRETE OVER STEEL FLOOR DECK

2 1/2"

EXTENDED END

2 1/2" MIN.

CEILING EXTENSION

HORIZONTAL BRIDGING ANGLES WELDED TO CHORDS

FIGURE 7.21 Open Web Joist Spans

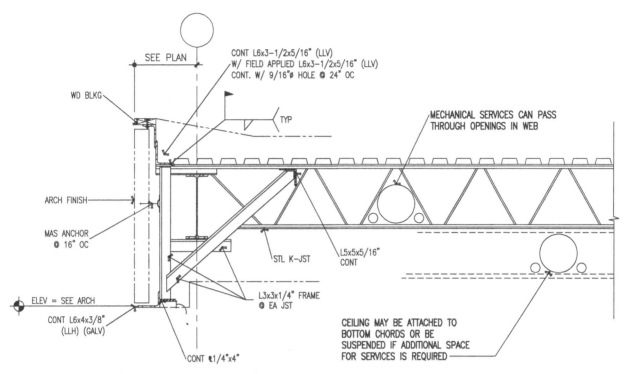

FIGURE 7.22 Utilities Supported from Joists

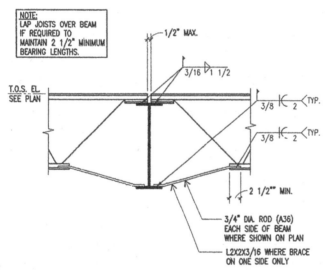

FIGURE 7.23 Welds for Supporting Joists on Beams and Girders

DIVISION 6, "WOODS, PLASTICS, AND COMPOSITES"

Wood frame construction is not used very often today in commercial construction, with the exception of retail or restaurant construction where wood materials allow the architect a lot of flexibility in design to satisfy the owner's desire for the facility to be unique in order to attract customers. As with masonry construction, wood framing can be used in a commercial building, usually after the piers, grade beams, and slab are

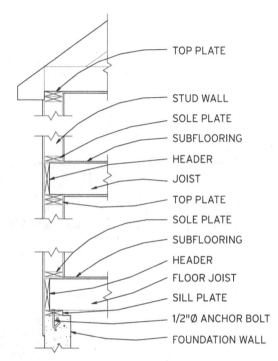

FIGURE 7.24 Attachment of Wood Framing to Grade Beams

in place. Figure 7.24 shows how the wood framing can be attached to the concrete grade beams using anchor bolts.

Structural wood framing is assembled with the same considerations for openings so that weight can be transferred around doors and windows. Additional wood members are used at openings and corners to provide the strength needed for holding the weight. Figure 7.25 shows typical framing for structural wood walls. Notice the construction methods for the headers so that weight can be transferred down the openings. Structural wood framing can cover multiple floors, as shown in Figure 7.26, and the weight transfers continually in a straight line down until it reaches the grade beams, the same as the masonry wall.

Structural wood framing can be used in combination with concrete, metal, and masonry structures. It can even be used to transition from those materials to a wood frame roof. In Figure 7.27 you can see samples of the types of wood joists that are available and how they can be used with the other systems. Notice their similarity in design to the metal open web joists.

The carpentry requirements for structural wood framing are beyond the scope of this text. In most cases in commercial construction where structural wood framing is used, the structural engineer will not design each member of the walls. He will note any specific requirements for the framing and the specifications will define the construction of the wall. It will typically be left to an experienced framing carpenter to construct each wall using the plans and specifications to meet the structural requirements of the building.

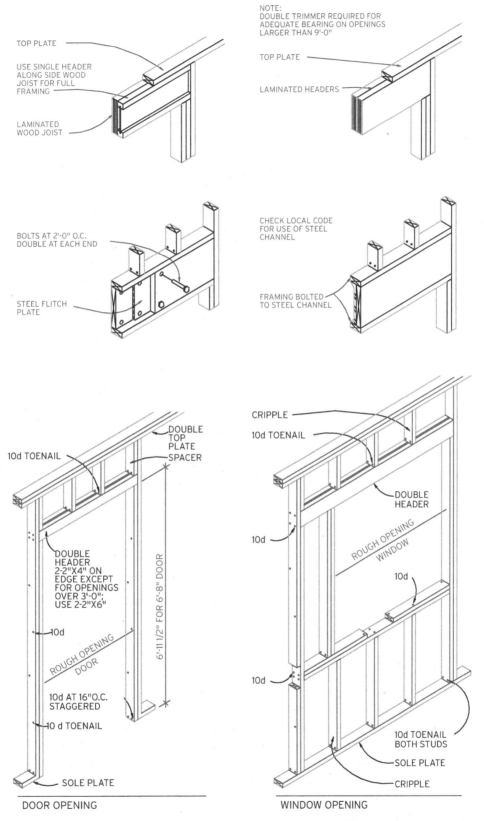

TOP PLATE

USE SINGLE HEADER ALONG SIDE WOOD JOIST FOR FULL FRAMING

LAMINATED WOOD JOIST

NOTE:
DOUBLE TRIMMER REQUIRED FOR ADEQUATE BEARING ON OPENINGS LARGER THAN 9'-0"

TOP PLATE

LAMINATED HEADERS

BOLTS AT 2'-0" O.C. DOUBLE AT EACH END

STEEL FLITCH PLATE

CHECK LOCAL CODE FOR USE OF STEEL CHANNEL

FRAMING BOLTED TO STEEL CHANNEL

10d TOENAIL

DOUBLE TOP PLATE

SPACER

DOUBLE HEADER 2-2"X4" ON EDGE EXCEPT FOR OPENINGS OVER 3'-0"; USE 2-2"X6"

6'-11 1/2" FOR 6'-8" DOOR

10d

ROUGH OPENING DOOR

10d AT 16"O.C. STAGGERED

10 d TOENAIL

SOLE PLATE

DOOR OPENING

CRIPPLE

10d TOENAIL

DOUBLE HEADER

10d

ROUGH OPENING WINDOW

10d

10d

10d TOENAIL BOTH STUDS

SOLE PLATE

CRIPPLE

WINDOW OPENING

FIGURE 7.25 Framing for Structural Wood Walls

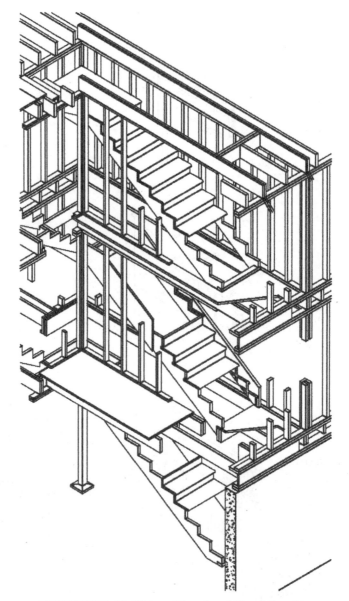

FIGURE 7.26 Multifloor Structural Wood Framing

SUMMARY

In reviewing structures, we should look at Figures 7.28 and 7.29, typical floor plans for the structure of a commercial building. Usually the architect and structural engineer will develop a plan view called the foundation framing plan, which will show the information necessary for the piers, grade beams and slab. Then, a plan view will be developed for the top portion of the structure, called the roof framing plan, which is used to show the beams, girders, and joists, or the top part of the structure.

Looking at Figure 7.28, at coordinates 20.5-D.7, notice the callout P2, for a Pier #2 at this location. The reader would locate the pier detail and

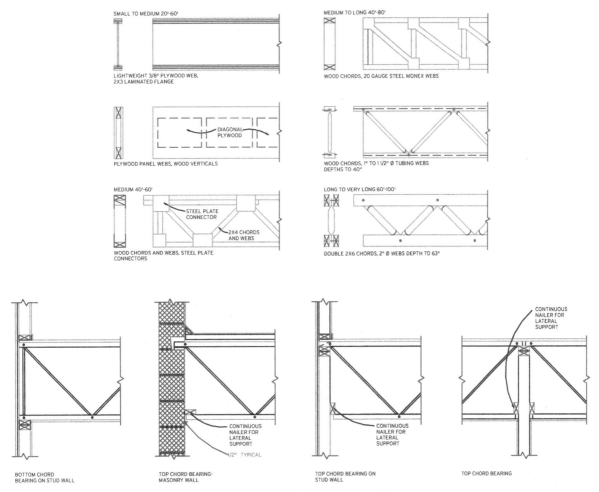

FIGURE 7.27 Wood Joist Examples

schedule in the structural drawings to find the definition of this pier. In addition, a C4 metal column goes at that location. Again, the reader would find the column details and schedules to obtain the definition of this structural element. At coordinates 18.2-E, notice the reference to the grade beam and the callout for Detail 3 on sheet S3.4 for that specific beam. Also notice the walls for this area are masonry, as indicated by the symbols shown in Figure 7.2.

In Figure 7.29 at coordinates 19.2 between E and E.2, we can see the callout for the joists that span the open area of the building, 32LH06. Per the nomenclature designations we learned earlier, these joists are 32 inches deep, are from the longspan series, and are the sixth type of joist used on this project. Between coordinates 19.2 and 19.4 at D.4, we can see that the outside supporting beam is a W10×12, meaning that it is a W Series beam, 10 inches deep, and weighs 12 pounds per linear foot. Note that C9×10.5 channels are used to span the distance from the wall to the W10×12 beam.

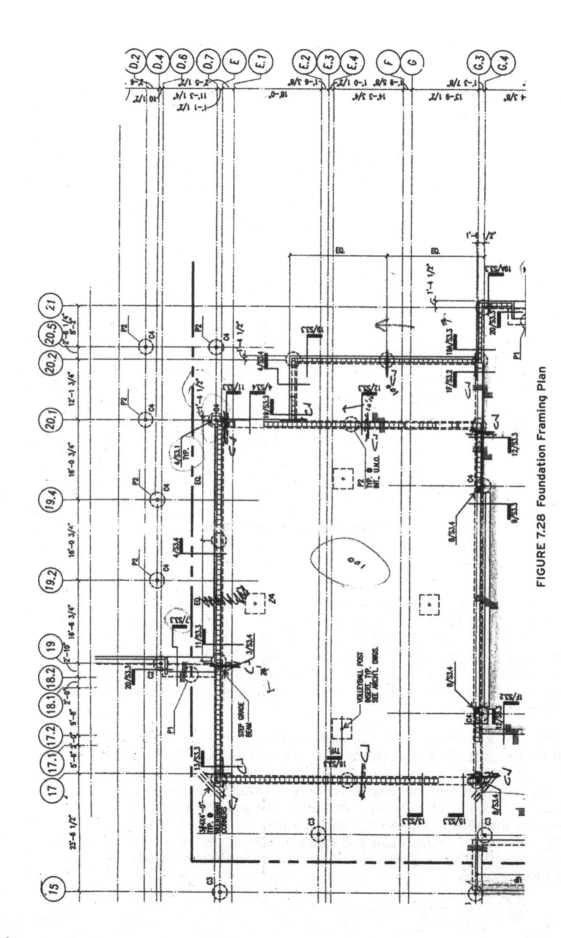

FIGURE 7.28 Foundation Framing Plan

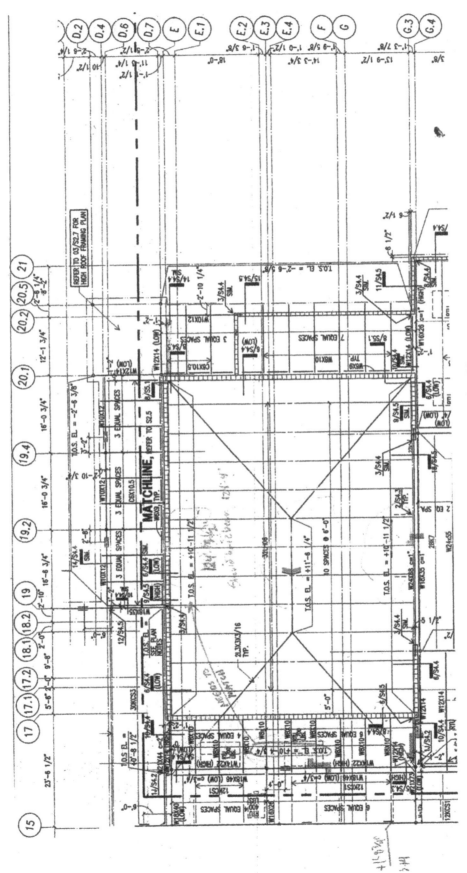

FIGURE 7.29 Roof Framing Plan

There are other features to note as well. See if you can find where the foundation plan shows changes in slab elevations, or where the roof framing plan shows the roof slope.

Now that we are somewhat familiar with the building envelope itself, we are ready to look the systems inside the building that allow it to function-primarily, the mechanical and electrical systems.

MECHANICAL CATEGORY OF THE DRAWINGS

OVERVIEW

The mechanical category of the drawings consists of information for several types of building systems that allow the building to be utilized and occupied by people. Although there can be many different mechanical systems, depending on what the building is being used for, in this book we are going to discuss how the more typical of these systems, HVAC, piping, and plumbing, are shown in the drawings. All of these systems could be combined into the single category called *mechanical*. However, in many sets of plans, these systems are so large and complex that they will be defined such that each is a separate drawing category.

These mechanical systems provide a suitable environment for conducting the business that the building was created for. The piping systems are used to convey various fluids to points of use within the building, and the plumbing systems are used to collect the fluids and divert them to collection systems. Although these are separate systems, they typically are linked to each other for complete operation of the building, as we will see later.

The architect will provide the mechanical engineer with the basic building drawings that were used in the architectural category, but without the dimensions and other data. The mechanical engineer will create plan views, using those as a background. The dimensions are removed to eliminate redundancy and to keep the mechanical system dimensions from being difficult to read. By using the architect's backgrounds for his drawings, the mechanical engineer saves time by not having to redraw

RELATED DIVISIONS

Division 15
(Old Format)

Division 23
(New Format)

Print and Specifications Reading for Construction, Updated Edition. Ron Russell.
© 2024 John Wiley & Sons, Inc. Published 2024 by John Wiley & Sons, Inc.
Companion website: www.wiley.com/go/printspecreadingupdatededition

the building and can more easily coordinate designs with the building elements. Now is a good time to recall the discussion we had in Chapter 6 regarding the reflected ceiling plan found in the architectural category. On this plan view, the suspended ceiling grid is laid out and the lighting, HVAC supply diffusers and return air grills, fire protection heads, and other systems that affect the ceiling—many of which are mechanical—are shown to coordinate their locations in the ceiling. In this discussion of mechanical systems, we are going to look strictly at HVAC systems. We will explore piping and plumbing in Chapter 10.

HVAC SYSTEMS

The HVAC systems in a commercial building can comprise several pieces of equipment or a single unit to accomplish all three tasks. The primary function of the HVAC systems is to heat or cool the air and distribute the conditioned air throughout the building space. If we look at Figure 8.1, we can see how each of these elements of an HVAC system works to accomplish this. Based on the figure, we can discuss the functions of an HVAC system so as to understand how each part of the system will be represented in the drawings.

Ventilating

Ventilating involves air movement, either to distribute conditioned air, to provide fresh air, or to remove contaminated air from the building or specific spaces or rooms. Most of these systems involve some type of conduit for the air to move in, such as ductwork, and have a fan of some type to create airflow. These systems will typically be fitted with an in-line filter system to clean the air.

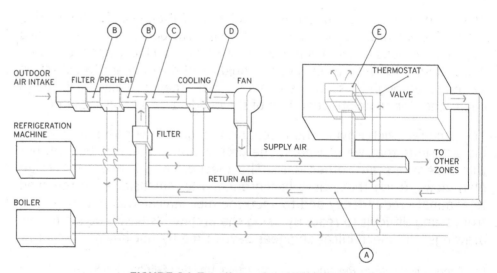

FIGURE 8.1 Functions of an HVAC System

AIR DISTRIBUTION

In Figure 8.1 we can see the air distribution system. This system starts with the fan unit, which would be sized to accommodate the amount of air that is required to be moved to maintain a certain temperature within the building spaces. This fan would typically be located somewhere centralized to the space it is supplying. It would be composed of an electric motor with a cage-type fan made of blades designed to move the air at a certain rate of speed, measured in cubic feet per minute (CFMs), to provide cooling or heat to a space. This is the supply side of the distribution system. As the supply air mixes with the existing air in the room, its temperature is modified and more air is required to maintain the desired temperature. However, as we force more supply air into the space, the air pressure builds up and must be relieved. We could release this air outside the building envelope, but since we have spent money for energy to heat or cool this air, we want to save the available energy left in it and minimize the cost of reheating or chilling new air from the outside. This requires a return air system that is designed to collect the air from the conditioned spaces and return it to the fan for reconditioning. Thus, the air is continually recirculated for efficiency in the HVAC system.

Because maintaining the air temperature is critical after the air has been conditioned, the air usually cannot be allowed to travel too great a distance, or the temperature will modify before reaching its destination and will not raise or lower the space temperature as needed. Therefore, the fan unit should be positioned as close as feasible to the spaces being conditioned. Sometimes this air distribution is accomplished using components called fan coil units (FCUs), E on Figure 8.1. These units function as a smaller-scale air distribution system. See Figure 8.2. In this system, FCU-25 is used to move the air into the room space through the main supply duct, which is 16"×16". This duct is shown using double-line drawing techniques because both sides

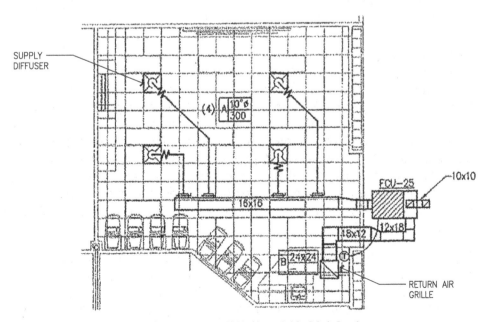

FIGURE 8.2 Fan Coil Unit and Air Dist System

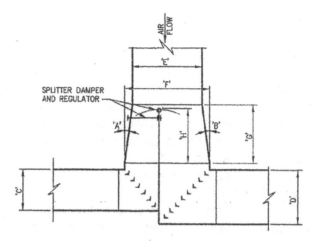

01 SPLITTER TEE DETAIL

SCALE: NOT TO SCALE

FIGURE 8.3 Duct Detail

of the duct are shown. Figure 8.3 shows a typical drawing detail of how this air distribution element would be constructed. The supply branches are shown as a single line, meaning that one line is used to show the entire duct. In Figure 8.3, supply branches are shown coming from the main supply duct and being routed to the supply diffusers through flex ducts so the air can be distributed into the space. Notice how the location of the diffusers is coordinated with the lights in the ceiling. This would be shown on the reflected ceiling plan as well. Notice the box in the middle of the room in Figure 8.2. This indicates that the four supply diffusers are to have 10-inch ducts run to them and that they should be balanced so that they supply 300 cubic feet of air per minute. Figure 8.4 shows a detail of how this is to be constructed. In the lower right of this figure (same room), there is a square that has a single slash across it. This is the symbol used to indicate the location of the return air grille. The box next to it shows that the return air duct should be 24″×24″; you can see how it is ducted back to the fan coil unit. This is a common method for achieving air distribution where a centralized fan is not the best option, and is representative of how the distribution elements are shown on the drawings.

Figure 8.5 shows a typical detail of how the fan coil unit would be suspended from the ceiling with the supply and return ducts attached to allow flow through the unit. Also note that there is a reference to an outside air intake duct. On the plan view shown in Figure 8.2, we can see that this duct is described as being 10″×10″. This duct allows us to provide additional air to the space from the outside, as needed. This is called make-up air.

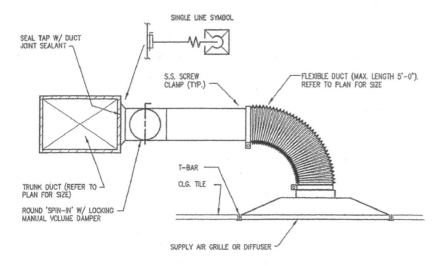

04 SPIN-IN & CLG. DIFFUSER DETAIL
SCALE: NOT TO SCALE

FIGURE 8.4 Diffuser Detail

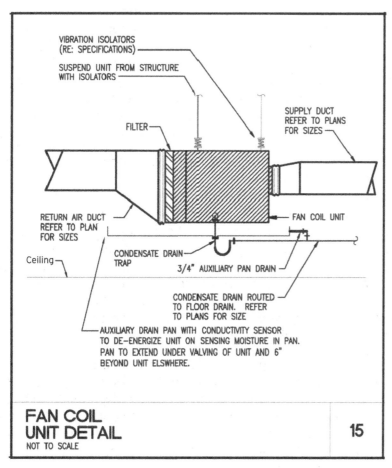

FIGURE 8.5 Fan Coil Detail

MAKE-UP AIR

Outside air is a source of air used to allow balancing of the air distribution system. Since most buildings cannot be made perfectly airtight, and it is not necessarily desirable to do so, we must have some mechanism for allowing additional air into the building in controllable amounts. Make-up air is sometimes called fresh-air make-up since it is allowing a source of fresh air into the building to make up for air that is lost through doors, windows, and other openings. In most municipalities, it is a requirement that a certain amount of fresh air is introduced into the building every hour so that the occupants will always have a certain percentage of fresh air in their building environment. In Figure 8.1 you can see that a duct B is extended outside the building envelope to allow fresh air to enter into the building's air distribution system. The entry point B1 is located so the outside air enters before the fan system so it can go through the preheating and cooling process C and D before being distributed to the building space. For FCU-25 shown in Figure 8.2, the outside air is supplied to the 10″×10″ ductwork using a separate outside air unit that would supply fresh make-up air to several FCUs. To maintain a balanced system, we also need to be able to relieve surplus air by using an exhaust system.

EXHAUST

Exhaust systems provide two functions in a commercial building. They can be used to remove air that has been contaminated by heat, cold, or dust from a local source or space, or they can be used to balance a larger air distribution system by relieving air to the outside. The exhaust air vent on an air duct system works the same as the one for make-up air, only the airflow is in the opposite direction. This exhaust vent allows excess air to be removed from the air distribution system so that the air flows in the correct volumes to the conditioned spaces and the system remains balanced.

Figure 8.6 shows a typical detail for exhaust systems that would be used for exhausting air from a specific local source that has been contaminated, or air that is not desirable for balancing the distribution system. This system would typically be used to exhaust air from an operation that creates contaminates that should not be allowed into the general room atmosphere and would be exhausted through a duct or hood collection system. Figures 8.7 and 8.8 show exhaust systems that would be used for a more generalized area, such as a kitchen area or shower/restroom area, where the conditions created in those areas would be annoying to the occupants if measures were not taken to change the atmosphere more frequently than general-occupancy areas.

Throughout the discussion of ventilation, we have talked about balancing the air distribution system. In Figure 8.9 we see a typical example of the methods used to determine what mix of supply, outside, exhaust, and make-up air is required to ensure that the correct amount of air is always present and circulating in a defined space. As air is removed, additional supply air must be provided. In the balancing schedule, we can see how the air sources are compared so that the room is balanced. Next, we need to see how the air being distributed is conditioned.

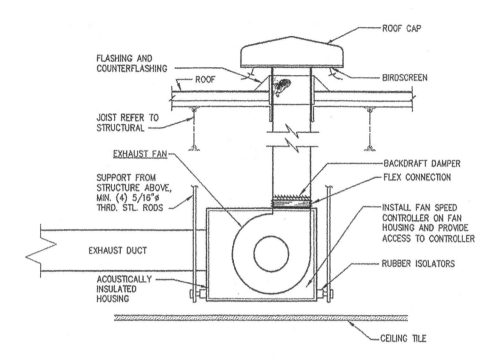

08 IN-LINE EXHAUST FAN DETAIL
SCALE: NOT TO SCALE

FIGURE 8.6 In-line Exhaust

Heating

Heating the air in the distribution system can be accomplished several ways, either by directly exposing the air to a heat source such as electrical coils, gas, or oil burners, or by using some intermediate source to carry the heat to the distribution system. In Figure 8.1 the illustration shows a typical method of providing heat to the air distribution system in a commercial building. Some type of fuel is fed into a boiler system to heat water, an intermediate source, which is pumped through a piping system to a coil inside the fan unit. As the air moves across the coils, the air absorbs the heat from the water and its temperature rises. The air distribution system then moves the heated air into the building spaces, where it gives its heat up to the objects in the room. The air is then recirculated back to the fan unit and across the coils to be reheated. If electric heat is being used, a series of electrical conductors would be used in place of the water coil inside the fan unit.

Air Conditioning (Cooling)

Cooling the air in the distribution system is usually accomplished by using some intermediate source to carry the heat removed from the air out of the building envelope. The different systems are identified by the intermediate medium used. Figure 8.1 shows a basic cooling system using chilled

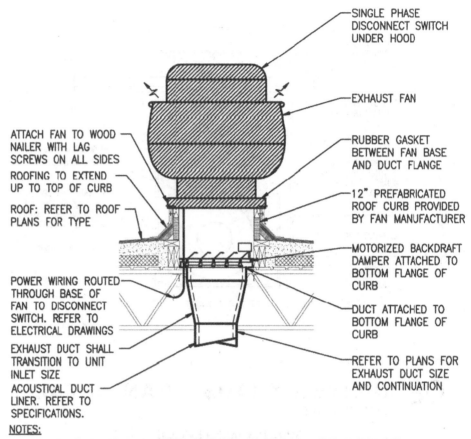

SINGLE PHASE
DISCONNECT SWITCH
UNDER HOOD

EXHAUST FAN

RUBBER GASKET
BETWEEN FAN BASE
AND DUCT FLANGE

12" PREFABRICATED
ROOF CURB PROVIDED
BY FAN MANUFACTURER

MOTORIZED BACKDRAFT
DAMPER ATTACHED TO
BOTTOM FLANGE OF
CURB

DUCT ATTACHED TO
BOTTOM FLANGE OF
CURB

REFER TO PLANS FOR
EXHAUST DUCT SIZE
AND CONTINUATION

ATTACH FAN TO WOOD
NAILER WITH LAG
SCREWS ON ALL SIDES

ROOFING TO EXTEND
UP TO TOP OF CURB

ROOF: REFER TO ROOF
PLANS FOR TYPE

POWER WIRING ROUTED
THROUGH BASE OF
FAN TO DISCONNECT
SWITCH. REFER TO
ELECTRICAL DRAWINGS

EXHAUST DUCT SHALL
TRANSITION TO UNIT
INLET SIZE

ACOUSTICAL DUCT
LINER. REFER TO
SPECIFICATIONS.

NOTES:

1. FOR THREE PHASE FANS, REFER TO THREE PHASE DISCONNECT SWITCH DETAIL.

2. PROVIDE DUCT LINER IN THE FIRST 10 FEET OF DUCT FROM FAN.

FIGURE 8.7 Exhaust System in a Kitchen

water. Water that has been cooled in the chiller is circulated through the coils in the cooling unit D, where heat in the air is absorbed into the water. The cooled air is circulated through the building spaces, where its temperature is modified, or increases, as it absorbs heat from the objects in the room, and it then returns through "A" to the fan unit–again, to give up its heat to the water in the coils. When the water returns to the chiller, it passes through another set of coils, where the heat is transferred to another medium, typically a refrigerant of some type in the form of a gas, thus lowering the temperature of the water, preparing it for its return to the coil in the fan unit to absorb more heat. The refrigerant then circulates to the cooling tower, or condensing unit, where fans circulate air through this unit's coils to remove the heat to the outside environment, causing the refrigerant to return to a liquid state. The refrigerant is then compressed, using a compressor, and returned to the chiller as a gas, where it is ready to remove more heat from the water.

In both the heating and cooling scenarios, intermediate mediums are used to heat and cool the air instead of in-line sources because there are great volumes of air to be conditioned, requiring large equipment to provide the required temperature differences. If additional air travel were to be

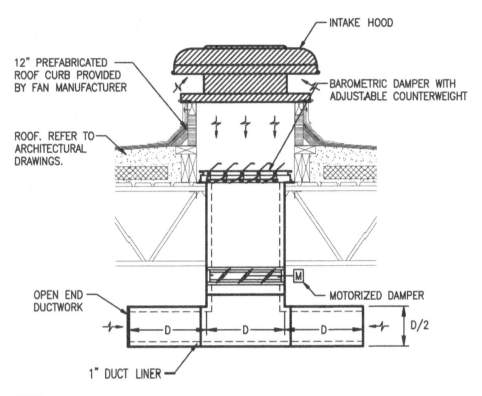

NOTES:
1. UNDER NO CIRCUMSTANCES SHOULD EQUIPMENT OF ANY KIND BE MOUNTED TO THE SIDE OF THE ROOF CURB.
2. FAN MANUFACTURER TO PROVIDE BAROMETRIC DAMPER AND ROOF CURB.
3. CONTROLS CONTRACTOR TO PROVIDE AND INSTALL CONTROL DAMPER WITH MOTORIZED DAMPER OPERATOR MOUNTED TO THE EXTERIOR OF DUCT.
4. MINIMUM DIMENSION OF CURB TO BE 12".

FIGURE 8.8 Exhaust System in a Shower/Restroom

required to receive heat or cooling directly from the source, the air would be modified significantly before reaching its destination within the building. How these intermediate sources are delivered to FCU-25 in Figure 8.2 will be discussed in Chapter 10. However, in those cases where the building space being conditioned is close to the fan source and the volumes are such that the distance of travel would not cause the air to modify significantly before being delivered to the space, another system can be used that contains all the equipment necessary to heat and cool the air, as well as provide make-up air. These systems are called rooftop units.

Rooftop Units

When the volume of air is not so great that it requires significant heating and cooling sources, the air in a space can be conditioned using a single unit that contains all of the elements previously discussed, including the heating and cooling sources, the fan to move the air, the compressor and condensing systems to process the refrigerant, and the ability to bring in fresh air from the outside. As indicated by the name, these units are typically set on the roof of the building. Figure 8.10 illustrates a schematic diagram of how these units function.

KITCHEN—CAFETERIA AIR BALANCE SCHEDULE						
UNIT MARK	SUPPLY AIR (CFM)	OUTSIDE AIR (CFM)	RETURN AIR (CFM)	HOOD EXHAUST	HOOD MAKE—UP	GENERAL EXHAUST
AHU—3	5000	800	4200			
AHU—4	5000	800	4200			
AHU—5	3500	1000	2500			
AHU—6	3500	1000	2500			
KSF—1					1800	
KSF—2					1800	
KEF—1				3000		
KEF—2				3000		
DEF—1						600
TEF—13						70
EF—14						120
EF—15						500
EF—16						500
TOTALS	17000	3600	13400	6000	3600	1790

```
MAKE—UP:
        RTU OUTSIDE AIR INTAKE                          3600
        FAN MAKE—UP                                     3600
                                                        ————
                                                        7200
EXHAUST:
        HOOD EXHAUST                                    6000
        GENERAL EXHAUST                                 1790
                                                        ————
                                                        7790
BALANCE RESULTS:
        MAKE—UP                                         7200
        EXHAUST                                         7790
                                              NEGATIVE  —590
```

FIGURE 8.9 Air Balance Schedule

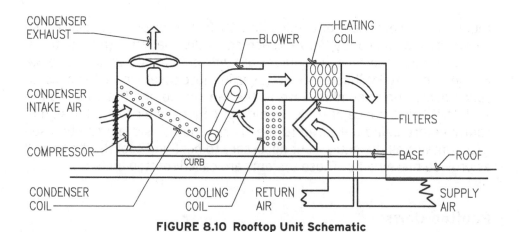

FIGURE 8.10 Rooftop Unit Schematic

In this example, we see the same elements that identified in Figure 8.1. The supply and return shows the air distribution system where the fan circulates the air out of the unit through the supply ductwork and draws the air back into the unit through the return ducts. The heating section consists of a series of either electrical heating elements or tubes that have natural gas fired through them and ignited to produce heat. As the air passes across these elements or tubes, it picks up the heat to deliver to the building space. The cooling system consists of a compressor to compress the refrigerant into a gas, which is then circulated through the cooling coils to pick up the heat from the air as it comes out of the building space. As the refrigerant heats up, it returns to a liquid state so that when it circulates through the condenser coils, the fan brings air across the coil, causing the refrigerant to give up its heat, which is then carried outside the unit in the condenser exhaust. The refrigerant then goes through the compressor to be compressed into a gas and be returned to the cooling coil to receive more heat from the building space. This one unit works for the entire building system shown in Figure 8.1, only on a much smaller scale.

Figure 8.11 shows a typical rooftop unit and the elements as they would appear installed. In the drawings, these pieces of equipment would be defined in schedules similar to those in Figure 8.12. In the plans, the locations of the RTUs must be coordinated with the structural drawings. The structural engineer will design certain elements of the building to support the weight of these units, based on the locations selected by the mechanical engineer. Several different configurations of HVAC systems are shown in Figure 8.11 and can be used depending on the size and configuration of the space being heated or cooled. Typical symbols used for the mechanical elements of the building are shown in Figure 8.13.

SUMMARY

This chapter discussed the air distribution system and provided some information about the media, or intermediate sources, used to transfer heat. We will discuss the conduits, or piping systems used to move these materials

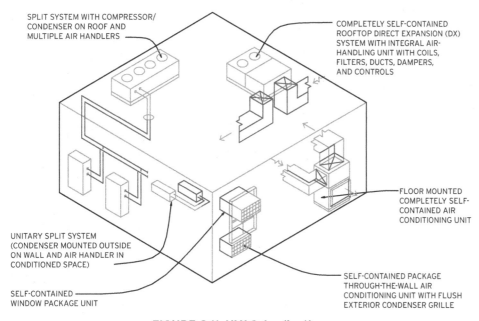

FIGURE 8.11 HVAC Applications

FAN SCHEDULE

DESIGNATION	LOCATION	SERVICE	MANUFACTURER	MODEL NUMBER	NOTES	FAN DATA TYPE	DRIVE	CFM	STATIC PRESSURE ("W.G.)	MOTOR HP (MIN.)	RPM (MAX.)
EF-01	ROOF	TOILET ROOM	COOK	120C13D	-	CENTRIFUGAL	DIRECT	1040	0.50	1/4	1,300
EF-02	ROOF	TOILET ROOM	COOK	120C13D	-	CENTRIFUGAL	DIRECT	1040	0.50	1/4	1,300
EF-03	ROOF	TOILET ROOM	COOK	135ACEB	-	CENTRIFUGAL	BELT	1190	0.50	FRAC	1,082
EF-04	ROOF	LOCKER RMTR	COOK	135C10D	-	CENTRIFUGAL	DIRECT	1460	0.50	1/2	1,216
EF-05	ROOF	COACH T/R	COOK	100ACEB	-	CENTRIFUGAL	BELT	360	0.25	FRAC	1,004
EF-06	ROOF	T/R / GYM	COOK	135ACEB	-	CENTRIFUGAL	BELT	940	0.25	FRAC	812
EF-07	ROOF	KILN ROOM	COOK	100ACEB	-	CENTRIFUGAL	BELT	400	0.25	FRAC	1,038
EF-08	ROOF	TOILET ROOM	COOK	120C13D	-	CENTRIFUGAL	DIRECT	1040	0.50	1/4	1,300
EF-09	ROOF	T/R/NURSE	COOK	100C3B	-	CENTRIFUGAL	BELT	700	0.25	1/4	1,725
EF-10	ROOF	KITCHEN T/R	COOK	80C2B	-	CENTRIFUGAL	BELT	390	0.25	1/6	1,725
EF-11	ROOF	T/R	COOK	100C2B	-	CENTRIFUGAL	BELT	590	0.25	1/6	1,725
EF-12	ROOF	T/R	COOK	80C2B	-	CENTRIFUGAL	BELT	440	0.25	1/6	1,725
EF-13	ROOF	BOYS LOCKER RMS	COOK	135C5B	-	CENTRIFUGAL	BELT	1860	0.25	1/2	1,725
EF-14	ROOF	GIRLS LOCKER RMS	COOK	135C4B	-	CENTRIFUGAL	BELT	1530	0.25	1/3	1,725
EF-15	ROOF	SCIENCE LAB	COOK	210R9B	-	CENTRIFUGAL	BELT	1600/3200	1.50	2	1,725
EF-16	ROOF	SCIENCE LAB	COOK	210R9B	-	CENTRIFUGAL	BELT	1600/3200	1.50	2	1,725
EF-17	ROOF	SCIENCE LAB	COOK	210R9B	-	CENTRIFUGAL	BELT	1600/3200	1.50	2	1,725
EF-18	ROOF	SCIENCE LAB	COOK	210R9B	-	CENTRIFUGAL	BELT	1600/3200	1.50	2	1,725
EF-19	ROOF	SCIENCE LAB	COOK	210R9B	-	CENTRIFUGAL	BELT	1600/3200	1.50	2	1,725
EF-20	ROOF	SCIENCE LAB	COOK	210R9B	-	CENTRIFUGAL	BELT	1600/3200	1.50	2	1,725
EF-21	ROOF	SCIENCE LAB	COOK	210R9B	-	CENTRIFUGAL	BELT	1600/3200	1.50	2	1,725
EF-22	ROOF	SCIENCE LAB	COOK	210R9B	-	CENTRIFUGAL	BELT	1600/3200	1.50	2	1,725
EF-23	ROOF	PREP ROOM	COOK	165 ACRLHP	-	CENTRIFUGAL	BELT	600/1200	1.50"	3/4	1,838
EF-24	ROOF	PREP ROOM	COOK	165 ACRLHP	-	CENTRIFUGAL	BELT	600/1200	1.50"	3/4	1,838
EF-25	ROOF	PREP ROOM	COOK	165 ACRLHP	-	CENTRIFUGAL	BELT	600/1200	1.50"	3/4	1,838
KSF-01	ROOF	KITCHEN	COOK	245VX10B	1-3, 8, 9	CENTRIFUGAL	BELT	3,000	3.00"	3	1,725
KSF-01	ROOF	KITCHEN	COOK	100KSP-B	1-3, 10	CABINET	BELT	1,800	1.50"	1	1,725
KSF-02	ROOF	KITCHEN	COOK	245VX10B	1-3, 8, 9	CENTRIFUGAL	BELT	3,000	3.00"	3	1,725
KSF-02	ROOF	KITCHEN	COOK	100KSP-B	1-3, 10	CABINET	BELT	1,800	1.50"	1	1,725
KSF-03	ROOF	KITCHEN	COOK	245VX10B	1-3, 8, 9	CABINET	BELT	3,000	3.00"	3	1,725
KSF-03	ROOF	KITCHEN	COOK	100KSP-B	1-3, 8, 9	CABINET	BELT	1,800	1.50"	1	1,725
KSF-04	ROOF	DISHWASHER	COOK	165RX7B	-	CENTRIFUGAL	BELT	1,200	2.00"	3/4	1,725
HEF-01	ROOF	LAB HOOD	COOK	165 ACRLHP	-	CENTRIFUGAL	BELT	1,600	1.50"	3/4	1,838

1. REFERENCE ELECTRICAL DRAWINGS FOR ELECTRICAL DATA.
2. REFERENCE SPECIFICATIONS FOR SEQUENCE OF OPERATIONS.
3. REFERENCE ARCHITECTURAL DRAWINGS FOR ROOF CURB DETAIL.

PACKAGE ROOF TOP UNIT WITH GAS HEAT SCHEDULE

DESIGNATION	SERVICE	TYPE	MANUFACTURER	MODEL NUMBER	ARRANGEMENT	NOTES	INDOOR BLOWER DATA TOTAL CFM	OUTSIDE AIR CFM	EST. EXT. SP. (IN. W.G.)	MOTOR H.P. (MIN.)	CFM OVER COIL	COOLING COIL DATA MAX. FACE VELOCITY (FPM)	GRAND SENSIBLE BTUH	GRAND TOTAL BTUH	EAT (°F DB)	EAT (°F WB)	LAT (°F DB)	GAS HEAT DATA FUEL	INPUT BTUH	OUTPUT BTUH	EAT (°F DB)	LAT (°F DB)
RTU-01	ADMIN	PACKAGED	TRANE	YHC063	HORIZONTAL	1-4	1,600	200	1.0	-	1,600	450	41,120	60,940	82.1°	69.4°	60.7°	N.G.	80000	66340	57.3°	94.6°
RTU-02	ADMIN	PACKAGED	TRANE	YHC072	HORIZONTAL	1-4	1,920	230	1.0	-	1,920	450	47,690	72,710	82.7°	69.9°	61.9°	N.G.	120000	99510	56.2°	103.3°
RTU-03	ADMIN	PACKAGED	TRANE	YHC092	HORIZONTAL	1-4	2,400	310	1.0	-	2,400	450	62,150	94,240	83.2°	70.3°	61.5°	N.G.	150000	124600	55.2°	102.3°

1. REFERENCE ELECTRICAL DRAWINGS FOR ELECTRICAL DATA.
2. ESTIMATED EXTERNAL STATIC PRESSURE INCLUDES LOSSES THROUGH DUCTWORK, AIR DEVICES, SOUND ATTENUATORS, ETC.
3. AIR HANDLING UNIT INTERNAL STATIC PRESSURE SHALL INCLUDE LOSSES THROUGH COILS, CASING, INTERNAL DAMPERS, AND 0.75" W.G. FOR DIRTY FILTERS.
4. FAN - FORWARD CURVED BLADES; 1,000 RPM MAXIMUM.

FIGURE 8.12 Rooftop Units and Exhaust Fans Schedules

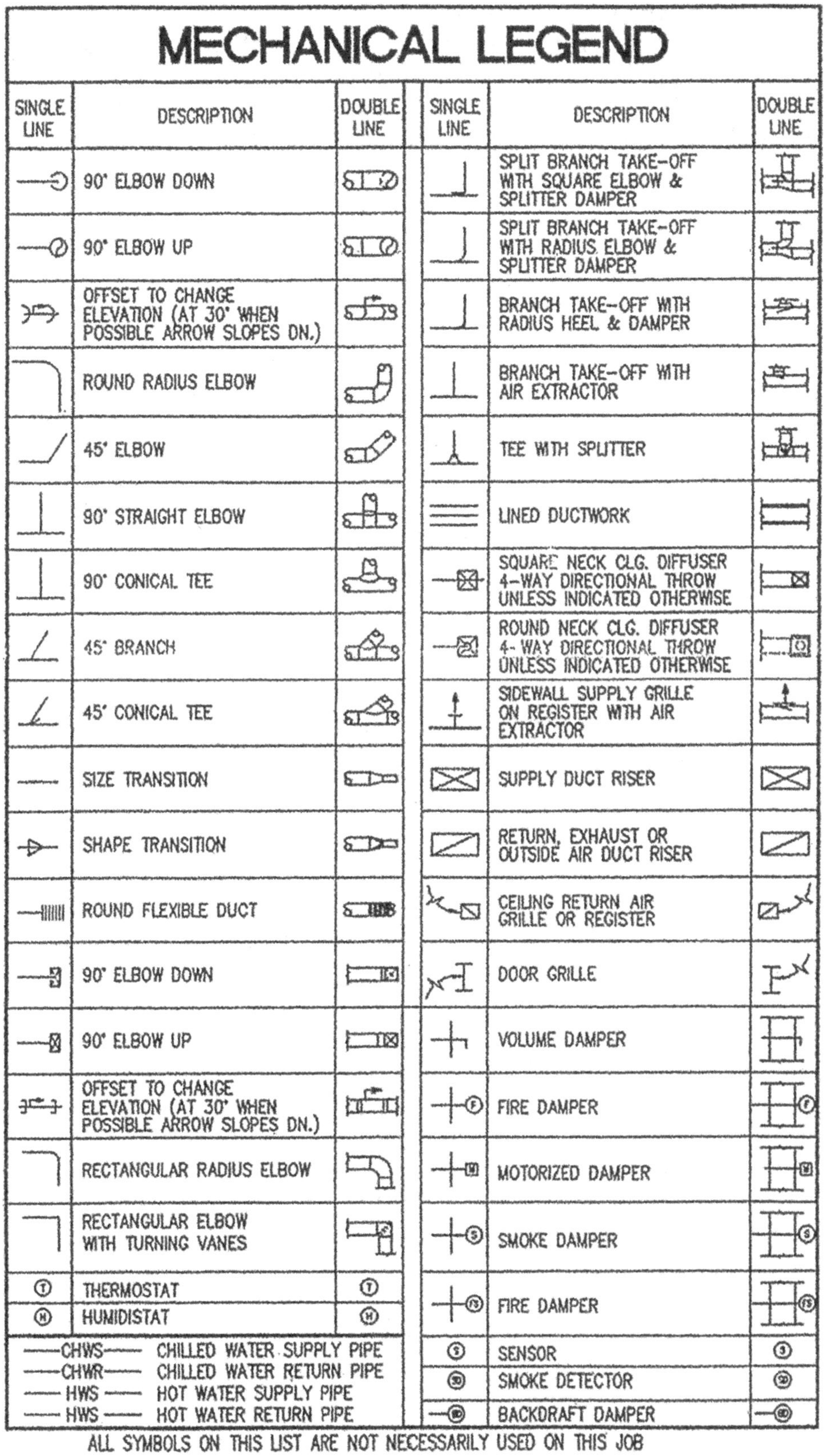

MECHANICAL LEGEND

SINGLE LINE	DESCRIPTION	DOUBLE LINE	SINGLE LINE	DESCRIPTION	DOUBLE LINE
	90° ELBOW DOWN			SPLIT BRANCH TAKE-OFF WITH SQUARE ELBOW & SPLITTER DAMPER	
	90° ELBOW UP			SPLIT BRANCH TAKE-OFF WITH RADIUS ELBOW & SPLITTER DAMPER	
	OFFSET TO CHANGE ELEVATION (AT 30° WHEN POSSIBLE ARROW SLOPES DN.)			BRANCH TAKE-OFF WITH RADIUS HEEL & DAMPER	
	ROUND RADIUS ELBOW			BRANCH TAKE-OFF WITH AIR EXTRACTOR	
	45° ELBOW			TEE WITH SPLITTER	
	90° STRAIGHT ELBOW			LINED DUCTWORK	
	90° CONICAL TEE			SQUARE NECK CLG. DIFFUSER 4-WAY DIRECTIONAL THROW UNLESS INDICATED OTHERWISE	
	45° BRANCH			ROUND NECK CLG. DIFFUSER 4-WAY DIRECTIONAL THROW UNLESS INDICATED OTHERWISE	
	45° CONICAL TEE			SIDEWALL SUPPLY GRILLE ON REGISTER WITH AIR EXTRACTOR	
	SIZE TRANSITION			SUPPLY DUCT RISER	
	SHAPE TRANSITION			RETURN, EXHAUST OR OUTSIDE AIR DUCT RISER	
	ROUND FLEXIBLE DUCT			CEILING RETURN AIR GRILLE OR REGISTER	
	90° ELBOW DOWN			DOOR GRILLE	
	90° ELBOW UP			VOLUME DAMPER	
	OFFSET TO CHANGE ELEVATION (AT 30° WHEN POSSIBLE ARROW SLOPES DN.)			FIRE DAMPER	
	RECTANGULAR RADIUS ELBOW			MOTORIZED DAMPER	
	RECTANGULAR ELBOW WITH TURNING VANES			SMOKE DAMPER	
(T)	THERMOSTAT	(T)		FIRE DAMPER	
(H)	HUMIDISTAT	(H)			
—CHWS—	CHILLED WATER SUPPLY PIPE		(S)	SENSOR	(S)
—CHWR—	CHILLED WATER RETURN PIPE		(SD)	SMOKE DETECTOR	(SD)
— HWS —	HOT WATER SUPPLY PIPE			BACKDRAFT DAMPER	
— HWS —	HOT WATER RETURN PIPE				

ALL SYMBOLS ON THIS LIST ARE NOT NECESSARILY USED ON THIS JOB

FIGURE 8.13 Mechanical Legend

in Chapter 10 when we discuss plumbing and piping. The equipment used will differ, depending on the size and configuration of the building and interior spaces and the processes being conducted at that facility. Most of this equipment will require electrical power, so now is a good time to discuss the electrical distribution system in a building. This is addressed in Chapter 9.

9

ELECTRICAL CATEGORY OF THE DRAWINGS

OVERVIEW

The electrical category of the drawings consists of information for several types of building electrical systems, including the power distribution system, lighting systems, and any communications systems, such as alarms, video, and fire detection systems, if required. It is possible that there are additional electrical systems, depending on the use of the building; however, we will focus on these common systems.

As with the mechanical drawings, the architect will provide the electrical engineer with the basic building drawings that were used in the architectural category—again, without the dimensions and other data. The electrical engineer will create plan views, using those as a background. Recall Chapter 6 and the reflected ceiling plan found in the architectural category of the drawings; these electrical elements are also shown on those drawings to coordinate their locations in the ceiling with the other ceiling-mounted items.

Even though the electrical drawings are structured similarly to the mechanical drawings, a different approach for examining them is recommended. One should first look at them to determine the overall main power distribution system, then the distribution of the power to the other systems such as lighting and communications. We will discuss this approach as we look at the types of information provided in the electrical drawings by the consulting electrical engineer.

RELATED DIVISIONS

Division 16 and 17
(Old Format)

Division 25, 26, 27, and 28 (New Format)

Print and Specifications Reading for Construction, Updated Edition. Ron Russell.
© 2024 John Wiley & Sons, Inc. Published 2024 by John Wiley & Sons, Inc.
Companion website: www.wiley.com/go/printspecreadingupdatededition

MAIN POWER DISTRIBUTION

To best understand the electrical system of a commercial building, we should understand how the consulting electrical engineer approaches designing these systems. The engineer will begin by identifying all of the power sources needed to power all of the building systems, equipment, lights, and any other items in the building that require power and then establish the design loads for each of them and for the entire building. Figure 9.1 shows a sample of the design load calculations for a small commercial building. These calculations will typically be included in the drawings so code officials and utilities reviewing the plans can see the engineer's requirements for the building. This will especially allow the utility company to determine the type and size of transformer needed for the building. In Figure 9.2 we can see how the three phases of power are provided from this transformer. These phases will allow the electrician to use single-, double-, and triple-pole breakers to contact each phase of power to run different devices. Once the electrical engineer has completed his load calculations, he is ready to begin developing the electrical distribution system for the building. The first drawing he will typically develop is the one-line diagram.

The one-line diagram, sometimes called electrical riser diagram, is a schematic drawing of the main electrical distribution system, starting with the entry of electrical service to the building, through the main switch gear, and out to the service panels. Figure 9.2 shows a simple electrical riser diagram. You can follow the distribution of electrical power from the transformer, through the main switchboard panel (MSB) and out to the different service panels DPHA, DPLA, CP, and KH and KL, AC1, H1, C1, L1, and so on, with H typically standing for high voltage and L representing low voltage. The panels are then numbered or lettered for what they are feeding, such as computer power, kitchen, or air conditioning.

Electrical Load Analysis		
PER NEC 220.34		
Total Connected Load	=	18.0 VA / Square Foot
School Square Footage	=	220,000 Square Feet
First 3 VA / Square Foot @ 100%	=	3 VA / Square Foot
Remaining 15.0 VA / Square Foot @75%	=	11.3 VA / Square Foot
Total Service Load	=	14.3 VA / Square Foot
Total Service KVA	=	3,135.0 KVA @ 480v, 3Ø
Total Ampacity	=	3,771 Amps
Service Ampacity	=	**4,000 Amps**

FIGURE 9.1 Electrical Load Analysis

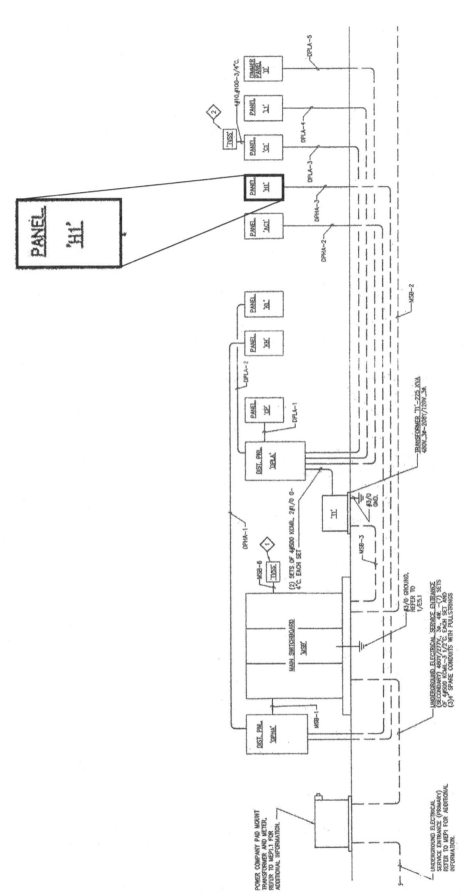

FIGURE 9.2 Electrical Riser Diagram

This drawing is called a one-line diagram because one line is used to represent the routing of the conduit and all wires from one device to another instead of having the drafter draw a line for each wire and conduit. The notes define what size conduit and wires are required and how many. The devices themselves are only shown to define the distribution -system—for example, the transformers that are shown. The transformers are indicated so the observer can easily determine when the voltage is changed to panels or devices. What is fed through or directly from each panel must be determined by looking at the panelboard schedule drawings. Figure 9.3 shows the description and the assignment of the switches for the main switchboard panel in the electrical riser diagram in Figure 9.2. From the MSB schedule, we can see that panel MSB is a service entrance rated panel and is rated for 2,500 amps. We can also see that circuit 1, or switch 1, feeds power to panel DPHA and that this switch is rated for 600 amps and is a three-pole switch. The feed from circuit 1 in panel MSB to panel DPHA consists of two (2) sets of 4#350KCMIL wires, plus a number 1 ground wire, all enclosed in a 3-inch conduit for each set. The panelboard schedule shown in Figure 9.4 provides similar information for the other panels indicated in the electrical riser diagram. There, we can see that panel H1 is rated for 200 amps at 277 or 480 V with all three phases of power. Panel DPHA contains three 200-amp three-pole breakers, seven

MAIN SWITCHBOARD "MSB" SCHEDULE

TYPE: SWITCHBOARD CONSTRUCTION: SERVICE ENTRANCE RATED   MAIN: 2500A. M.C.B. *

VOLTS: 480Y/277.,3∅.,4W.   BRACED: 65 KAIC

CCT.	SERVES	LOAD (KVA)	FRAME	P	TRIP	FEEDER
1	DIST. PANEL "DPHA"	311.0	600	3	600	(2)SETS OF 4#350KCMIL, #1G.–3"C. EACH SET
2	DIST. PANEL "DPHB"	590.0	1200	3	1200 *	(4)SETS OF 4#500 KCMIL, #3/0G.–3 1/2"C. EACH SET
3	DIST. PANEL "DPLA" VIA TRANSFORMER "T1"	201.0	400	3	350	3#500 KCMIL, #3G.–3"C.
4	WCU–1	192.0	400	3	300	3#350 KCMIL, #3G.–3"C.
5	WCU–2	192.0	400	3	300	3#350 KCMIL, #3G.–3"C.
6	TVSS	–	100	3	60	4#6, #10G–1"C.
7	SPARE	–	400	3	400	
8	SPACE	–	400	3	–	
9	SPARE	–	200	3	200	
10	SPACE	–	200	3	–	
	TOTAL	1486.0				

NOTE: IF ALTERNATE NO. 5 IS ACCEPTED, TRIP FOR WCU–1 AND 2 SHALL BE REDUCED TO 250 AMPS WITH 3#250 KCMIL, #4G–2 1/2"C.

* PROVIDE GROUND FAULT PROTECTION.

FIGURE 9.3 Main Panel/Switchboard Schedule

DISTRIBUTION PANEL "DPHA"

TYPE:	PANELBOARD CONSTRUCTION			MAIN:	600 A. M.L.O.	
VOLTS:	480Y/277V.,3∅.,4W.			BRACED:	35 KAIC	

CCT.	SERVES	LOAD (KVA)	FRAME	P	TRIP	FEEDER
1	PANEL "KH"	101.0	200	3	200	4#4/0,#6G.–2 1/2"C.
2	PANEL "AC1"	56.0	100	3	100	4#1,#6G–1 1/2"C
3	PANEL "H1"	68.0	200	3	200	4#3/0,#6G.–2"C.
4	CHWP–1	22.0	100	3	70	3#4,10G–1"C
5	CHWP–2	22.0	100	3	70	3#4,10G–1"C
6	HWP–1	17.0	100	3	60	3#4,10G–1"C
7	HWP–2	17.0	100	3	60	3#4,10G–1"C
8	UH–1	8.0	100	3	20	3#12,#12G.–1/2"C.
9	SPARE	–	200	3	200	–
10	SPACE	–	200	3	–	–
11	SPARE	–	100	3	100	–
12	SPACE	–	100	3	–	–
13	SPACE	–	200	3	–	–
14	SPACE	–	200	3	–	–
	TOTAL	311.0				

FIGURE 9.4 Distribution Panel Schedule

three-pole breakers, and spaces for four more three-pole breakers. Using these schedules, the electrician can understand the basic distribution of power within the building.

The electrical riser diagram only shows the sequence of the electrical devices that make up the major components of the building electrical system. It does not show where these components are located within the building.

Now that we have some understanding of the main power distribution system for the building, it is necessary that we know where these devices are located within the building. The electrical riser diagram shows panel DPHA in the schematic but does not indicate its physical location in the building. Looking at Figure 9.5, we can see in the lower-right corner of the plan view that panel MSB is located on an interior wall in an electrical room and that panel DPHA is located next to it. Also note that transformer T1 and panel DPLA are also located in this room across from MSB. If we look at coordinate location H16, we will see the location of the electrical room where panels AC1, H1, C1, and L1 are located. We now know where each of the main distribution devices is located in the building.

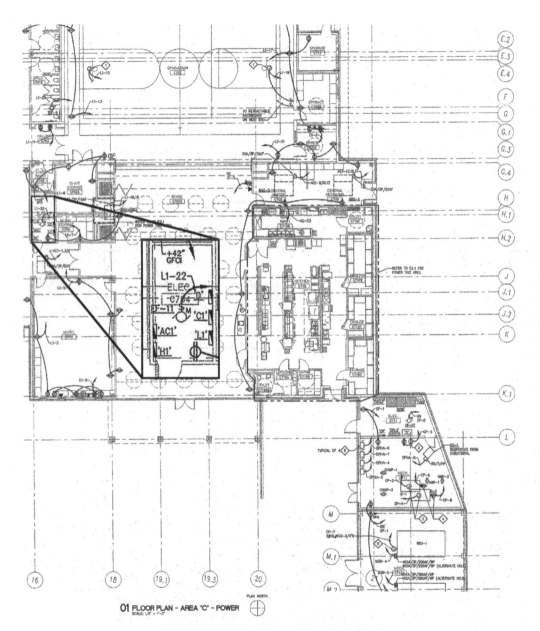

01 FLOOR PLAN – AREA 'C' – POWER
SCALE: 1/8" = 1'-0"

FIGURE 9.5 First-Floor Electrical Plan

Using this plan view and the electrical riser diagram information, the electrician can calculate quantities of materials that are needed to run the conduit and wire to each device. We know from the electrical riser diagram that transformer T1 is fed from panel MSB on circuit 3. We also know from the panel schedule in Figure 9.3 that circuit 3 requires three number 500KCMIL wires and a single number 3 ground wire, and that they are run in a 3-inch conduit. Seeing where transformer T1 is located on the plan view now allows us to determine what lengths of these materials are required to complete the circuit. Also note that, like the mechanical drawings, the electrical plan views do not show a specific routing to be taken for the conduit and wires. The coordination of the exact routing is left to the electrician and the general contractor. They can choose the routing that best meets the purpose of the overall installation of this and other crafts' work.

SECONDARY LEVEL OF POWER DISTRIBUTION

Now that we understand the main power distribution system for the building, we need to understand the next level of distribution and where the components are located. To understand this, we will look at the plan view in Figure 9.5 again. We understand now where panel H1 is located but need to know what is fed from this panel and where the devices are located. Let's start with the electrical for the lighting for the gym in Figure 9.6.

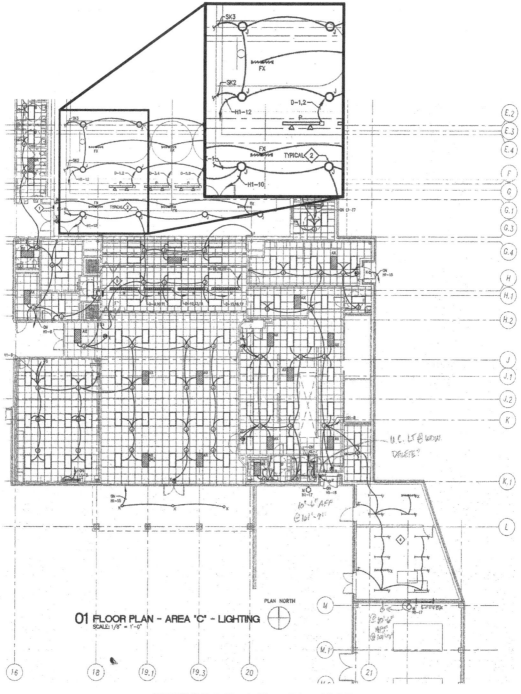

FIGURE 9.6 First-Floor Plan—Lighting

The gym and its lighting are shown at coordinates G18. There we see the symbol for lighting and an arrow that is labeled H1-12. This indicates that this circuit for this lighting circuit is from panel H1 and is on breaker number 12. Again, no specific routing of conduits is suggested to allow the electrician to select the most favorable route from the lighting location to panel H1. As you look at the rest of the plan views, Figures 9.5 and 9.6, you will notice that all of the circuits from each panel will be labeled this way. Also, notice that as in residential, many of the convenience outlets will be included on the same circuit since it is not anticipated that they will all have something plugged into them and operating at the same time. However, look at the circuit indicated between G.4 and H along 18.

The circuit indicated here shows an electrical symbol indicating a dedicated circuit, and the arrow indicating the circuit is fed from panel C1, breaker 3. This circuit is obviously designed for use of a single device and is dedicated to it since no other devices are indicated by connection to the arrow. To understand what the actual device is that will be on this circuit, one would most likely have to refer back to the first-floor plan in the architectural category of drawings.

As a way for the electrician to understand each of the circuits in a panel without having to scour the plan views, the panelboard schedule shown in Figure 9.7 for panel H1 indicates each circuit, its rating, how many poles it uses, and the name and location of the device. The panel rating for amps and voltages, as well as the loads calculated by the electrical engineer, are also shown. If we look at the breaker number 12, we can see that it services the gym lights, as we saw earlier. The electrician can use this schedule to premake the panel before installation. The electrician will construct the panel using the type of enclosure specified and build the bussing necessary to provide the poles needed and will purchase and install the needed breakers. This schedule is also used to label the actual panels so that people occupying the building will know what breakers operate what devices.

We have looked at a device shown on an electrical plan view, the outlet or receptacle. This is only an example of the many electrical devices that could be used in a commercial building. Figure 9.8 shows some of the many other devices that might be used on a commercial building project. As with all schedules like this, each one should be closely evaluated to determine which symbols represent which devices and which devices are used on the project you are working on. It is also important to note that electrical circuits for some major pieces of building equipment could also be identified with a schedule. In Figure 9.9 the circuits for the HVAC rooftop units are specified by schedule. RTU-A100 is fed from panel HAB, circuits 1, 3, and 5, which is a three-phase circuit. This would be the information provided to the electrician who would have to determine where RTU-A100 and panel HAB are located, and determine the best routing of the conduit and wires.

PANELBOARD 'H1'
480Y/277; 3-PHASE/4-WIRE
WITH EQUIPMENT GROUND BUS

225 A MLO
225 MAIN BUS
100% NEUTRAL BUS

LOCATION: ELEC ROOM C704
MOUNT: SURFACE
TYPE NEMA 1

18 KAIC BUS AND BRACING

L	MOTOR	OTHER	C.B.	POLE	LOAD AREAS AND LOCATION	A B C	LOAD AREAS AND LOCATION	POLE	C.B.	OTHER	MOTOR	L
3960			20/1	1	LIGHTS – B901,B902 AND MRS		LIGHTS – B603,B604 CORRIDOR &N MRS	2	20/1			3500
3300			20/1	3	LIGHTS – B805,B806 AND B807		LIGHTS – B509,B601,B608 & CORRIDOR	4	20/1			3025
3300			20/1	5	LIGHTS – B801,B802 AND B803		LIGHTS – CORRIDORS,RESTROOMS & MRS	6	20/1			3685
3355			20/1	7	LIGHTS – B115,CORRIDOR AND MRS		LIGHTS – CAFETERIA & MUSIC ROOM	8	20/1			3630
3355			20/1	9	LIGHTS – B502,B503 AND B504		LIGHTS – GYM & STAGE	10	20/1			3855
3300			20/1	11	LIGHTS – B505,B506 & CORRIDOR		LIGHTS – GYM	12	20/1			3680
3190			20/1	13	LIGHTS – B805, B806 AND B807		LIGHTS GYM & MR'S	14	20/1			2390
1520			20/1	15	CANOPY LIGHTS		PARKING LOT LIGHTS	16	20/2			1840
900			20/1	17	SECURITY LIGHTS			18				1840
			20/1	19	SPARE		SPARE	20	20/1			
			20/1	21	SPARE		SPARE	22	20/1			
			20/1	23	SPARE		SPARE	24	20/1			
			20/1	25	SPARE		SPACE	26				
			20/1	27	SPARE		SPACE	28				
			20/1	29	SPARE		SPACE	30				
				31	SPACE		SPACE	32				
				33	SPACE		SPACE	34				
				35	SPACE		SPACE	36				
				37	SPACE		SPACE	38				
				39	SPACE		SPACE	40				
				41	SPACE		SPACE	42				

CONNECTED KVA 3Ø		DEMAND/FACTOR		AMPS
LIGHTING	53.2	100%	53.2	64.0
MOTOR	0.0	N/A	0.0	0.0
OTHER	0.0	N/A	0.0	0.0
SPARE	15.0	100%	15.0	18.0
TOTAL	68.2		68.2	82.1

FEEDER AMPACITY ADJUSTMENTS
LIGHT AMPS   15.0
MOTOR AMPS   0.0

REMARKS: VA ENERGY MANAGEMENT SYSTEM

CONNECTED 3Ø LOAD CALCULATIONS

TOTAL KVA (ØA) =	71.5	AMPS
TOTAL KVA (ØB) =	60.2	AMPS
TOTAL KVA (ØC) =	60.3	AMPS

19.8
19.7
18.7
84.0   TOTAL CONNECTED 3Ø AMPS
11.3   NEUTRAL AMPS
18%   IMBALANCE

98 MINIMUM FEEDER AMPS

PER NEC 220-11
PER NEC 220-18 thru 41

FIGURE 9.7 Panel Schedule

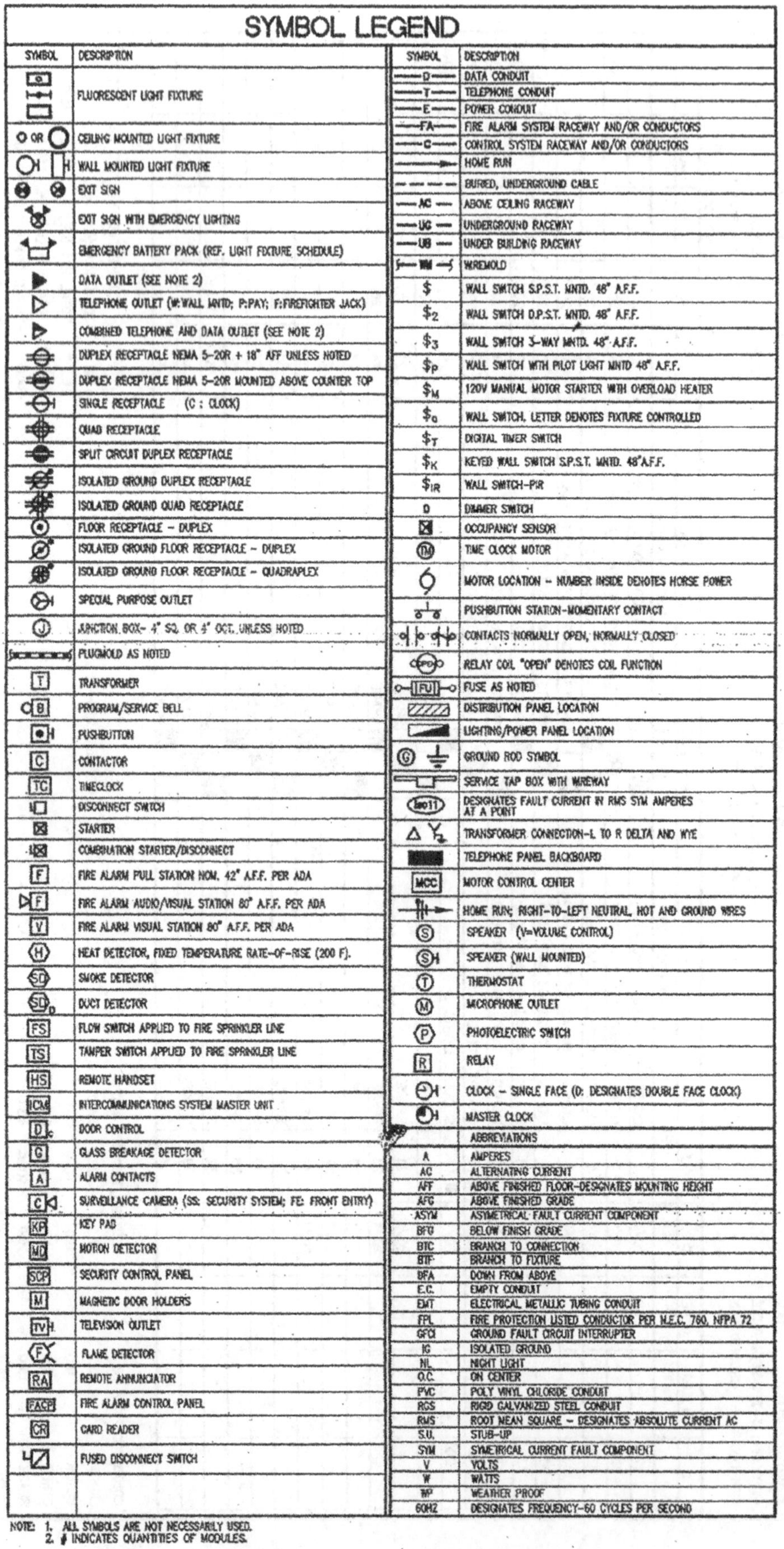

SYMBOL LEGEND

SYMBOL	DESCRIPTION	SYMBOL	DESCRIPTION
	FLUORESCENT LIGHT FIXTURE	—D—	DATA CONDUIT
		—T—	TELEPHONE CONDUIT
		—E—	POWER CONDUIT
O OR O	CEILING MOUNTED LIGHT FIXTURE	—FA—	FIRE ALARM SYSTEM RACEWAY AND/OR CONDUCTORS
		—C—	CONTROL SYSTEM RACEWAY AND/OR CONDUCTORS
	WALL MOUNTED LIGHT FIXTURE		HOME RUN
	EXIT SIGN	— — —	BURIED, UNDERGROUND CABLE
	EXIT SIGN WITH EMERGENCY LIGHTING	—AC—	ABOVE CEILING RACEWAY
		—UG—	UNDERGROUND RACEWAY
	EMERGENCY BATTERY PACK (REF. LIGHT FIXTURE SCHEDULE)	—UB—	UNDER BUILDING RACEWAY
		—WM—	WIREMOLD
▶	DATA OUTLET (SEE NOTE 2)	$	WALL SWITCH S.P.S.T. MNTD. 48" A.F.F.
▷	TELEPHONE OUTLET (W:WALL MNTD; P:PAY; F:FIREFIGHTER JACK)	$₂	WALL SWITCH D.P.S.T. MNTD. 48" A.F.F.
▶	COMBINED TELEPHONE AND DATA OUTLET (SEE NOTE 2)	$₃	WALL SWITCH 3-WAY MNTD. 48" A.F.F.
	DUPLEX RECEPTACLE NEMA 5-20R + 18" AFF UNLESS NOTED	$ₚ	WALL SWITCH WITH PILOT LIGHT MNTD 48" A.F.F.
	DUPLEX RECEPTACLE NEMA 5-20R MOUNTED ABOVE COUNTER TOP	$ₘ	120V MANUAL MOTOR STARTER WITH OVERLOAD HEATER
	SINGLE RECEPTACLE (C : CLOCK)	$ₐ	WALL SWITCH, LETTER DENOTES FIXTURE CONTROLLED
	QUAD RECEPTACLE	$ₜ	DIGITAL TIMER SWITCH
	SPLIT CIRCUIT DUPLEX RECEPTACLE	$ₖ	KEYED WALL SWITCH S.P.S.T. MNTD. 48"A.F.F.
	ISOLATED GROUND DUPLEX RECEPTACLE	$ᵢᵣ	WALL SWITCH–PIR
	ISOLATED GROUND QUAD RECEPTACLE	D	DIMMER SWITCH
	FLOOR RECEPTACLE – DUPLEX	⊠	OCCUPANCY SENSOR
	ISOLATED GROUND FLOOR RECEPTACLE – DUPLEX	⊕	TIME CLOCK MOTOR
	ISOLATED GROUND FLOOR RECEPTACLE – QUADRAPLEX		MOTOR LOCATION – NUMBER INSIDE DENOTES HORSE POWER
	SPECIAL PURPOSE OUTLET		PUSHBUTTON STATION-MOMENTARY CONTACT
J	JUNCTION BOX– 4" SQ. OR 4" OCT. UNLESS NOTED		CONTACTS NORMALLY OPEN, NORMALLY CLOSED
	PLUGMOLD AS NOTED		RELAY COIL "OPEN" DENOTES COIL FUNCTION
T	TRANSFORMER	—[FU]—	FUSE AS NOTED
B	PROGRAM/SERVICE BELL		DISTRIBUTION PANEL LOCATION
	PUSHBUTTON		LIGHTING/POWER PANEL LOCATION
C	CONTACTOR	G	GROUND ROD SYMBOL
TC	TIMECLOCK		SERVICE TAP BOX WITH WIREWAY
	DISCONNECT SWITCH		DESIGNATES FAULT CURRENT IN RMS SYM AMPERES AT A POINT
⊠	STARTER	△ Y	TRANSFORMER CONNECTION–L TO R DELTA AND WYE
	COMBINATION STARTER/DISCONNECT		TELEPHONE PANEL BACKBOARD
F	FIRE ALARM PULL STATION NOM. 42" A.F.F. PER ADA	MCC	MOTOR CONTROL CENTER
F	FIRE ALARM AUDIO/VISUAL STATION 80" A.F.F. PER ADA		HOME RUN; RIGHT-TO-LEFT NEUTRAL, HOT AND GROUND WIRES
V	FIRE ALARM VISUAL STATION 80" A.F.F. PER ADA	S	SPEAKER (V=VOLUME CONTROL)
H	HEAT DETECTOR, FIXED TEMPERATURE RATE-OF-RISE (200 F).	S	SPEAKER (WALL MOUNTED)
SD	SMOKE DETECTOR	T	THERMOSTAT
SD_D	DUCT DETECTOR	M	MICROPHONE OUTLET
FS	FLOW SWITCH APPLIED TO FIRE SPRINKLER LINE	P	PHOTOELECTRIC SWITCH
TS	TAMPER SWITCH APPLIED TO FIRE SPRINKLER LINE	R	RELAY
HS	REMOTE HANDSET		CLOCK – SINGLE FACE (D: DESIGNATES DOUBLE FACE CLOCK)
ICM	INTERCOMMUNICATIONS SYSTEM MASTER UNIT		MASTER CLOCK
D_c	DOOR CONTROL		**ABBREVIATIONS**
G	GLASS BREAKAGE DETECTOR	A	AMPERES
A	ALARM CONTACTS	AC	ALTERNATING CURRENT
C	SURVEILLANCE CAMERA (SS: SECURITY SYSTEM; FE: FRONT ENTRY)	AFF	ABOVE FINISHED FLOOR–DESIGNATES MOUNTING HEIGHT
KP	KEY PAD	AFG	ABOVE FINISHED GRADE
		ASYM	ASYMETRICAL FAULT CURRENT COMPONENT
MD	MOTION DETECTOR	BFG	BELOW FINISH GRADE
SCP	SECURITY CONTROL PANEL	BTC	BRANCH TO CONNECTION
		BTF	BRANCH TO FIXTURE
M	MAGNETIC DOOR HOLDERS	BFA	DOWN FROM ABOVE
TV	TELEVISION OUTLET	E.C.	EMPTY CONDUIT
FX	FLAME DETECTOR	EMT	ELECTRICAL METALLIC TUBING CONDUIT
		FPL	FIRE PROTECTION LISTED CONDUCTOR PER N.E.C. 760, NFPA 72
RA	REMOTE ANNUNCIATOR	GFCI	GROUND FAULT CIRCUIT INTERRUPTER
FACP	FIRE ALARM CONTROL PANEL	IG	ISOLATED GROUND
CR	CARD READER	NL	NIGHT LIGHT
		O.C.	ON CENTER
	FUSED DISCONNECT SWITCH	PVC	POLY VINYL CHLORIDE CONDUIT
		RGS	RIGID GALVANIZED STEEL CONDUIT
		RMS	ROOT MEAN SQUARE – DESIGNATES ABSOLUTE CURRENT AC
		S.U.	STUB-UP
		SYM	SYMETRICAL CURRENT FAULT COMPONENT
		V	VOLTS
		W	WATTS
		WP	WEATHER PROOF
		60HZ	DESIGNATES FREQUENCY–60 CYCLES PER SECOND

NOTE: 1. ALL SYMBOLS ARE NOT NECESSARILY USED.
2. # INDICATES QUANTITIES OF MODULES.

FIGURE 9.8 Electrical Symbol Legend

PACKAGE ROOF TOP UNIT WITH GAS HEAT SCHEDULE

DESIGNATION	SERVICE	VOLT/Ø/HZ	UNIT MCA	UNIT MOCP	OPD	CIRCUIT
RTU-A100	ADMINISTRATION	460/3/60	29	35	60/3/35F/WP	HAB-1/3/5
RTU-A104	ADMINISTRATION	460/3/60	19	25	30/3/25F/WP	HAB-7/9/11
RTU-A117	ADMINISTRATION	460/3/60	16	20	30/3/20F/WP	HAB-13/15/17
RTU-A121A	DINING ROOM	460/3/60	42	50	60/3/50F/WP	HAB-19/21/23
RTU-A121B	DINING ROOM	460/3/60	42	50	60/3/50F/WP	HAB-25/27/29
RTU-A122	STAGE	460/3/60	19	25	30/3/25F/WP	HAB-31/33/35
RTU-A123	LIBRARY	460/3/60	39	50	60/3/50F/WP	HAB-37/39/41
RTU-A201	COMPUTER SCIENCE	460/3/60	19	25	30/3/25F/WP	HAB-2/4/6
RTU-A202	ACE	460/3/60	19	25	30/3/25F/WP	HAB-8/10/12
RTU-A203	MATH	460/3/60	16	20	30/3/20F/WP	HAB-14/16/18
RTU-A204	MATH	460/3/60	16	20	30/3/20F/WP	HAB-20/22/24
RTU-B102	WORK ROOM	460/3/60	14	20	30/3/20F/WP	HDB-1/3/5
RTU-B104	MEDIA STR.	460/3/60	26	30	30/3/30F/WP	HDB-7/9/11
RTU-B113A	SERVING	460/3/60	39	50	60/3/50F/WP	HDB-13/15/17
RTU-B113B	SERVING	460/3/60	24	30	30/3/30F/WP	HDB-19/21/23
RTU-B113C	SERVING	460/3/60	39	50	60/3/50F/WP	HDB-25/27/29
RTU-C105	AP SEC.	460/3/60	26	30	30/3/30F/WP	HAB-26/28/30
RTU-C110	AP PRINT	460/3/60	19	25	30/3/25F/WP	HAB-32/34/36
RTU-C120	PRE-EMPLOYMENT	460/3/60	19	25	30/3/25F/WP	HAB-38/40/42
RTU-C121	CONTENT MASTERY	460/3/60	19	25	30/3/25F/WP	HAB-44/46/48
RTU-C122	LIFE SKILLS	460/3/60	24	30	30/3/30F/WP	HAB-50/52/54
RTU-D100	HEALTH	460/3/60	16	20	30/3/20F/WP	HDB-31/33/35
RTU-D102	SOCIAL STUDIES	460/3/60	16	20	30/3/20F/WP	HDB-37/39/41
RTU-D103	SPEECH	460/3/60	12	15	30/3/15F/WP	HDB-2/4/6
RTU-D104	SOCIAL STUDIES	460/3/60	16	20	30/3/20F/WP	HDB-8/10/12
RTU-D105	RESOURCE	460/3/60	16	20	30/3/20F/WP	HDB-14/16/18
RTU-D106	SOCIAL STUDIES	460/3/60	19	25	30/3/25F/WP	HDB-20/22/24
RTU-D110	SOCIAL STUDIES	460/3/60	16	20	30/3/20F/WP	HDB-26/28/30
RTU-D112	SOCIAL STUDIES	460/3/60	16	20	30/3/20F/WP	HDB-32/34/36
RTU-D113	SOCIAL STUDIES	460/3/60	16	20	30/3/20F/WP	HDB-38/40/42
RTU-D115	SOCIAL STUDIES	460/3/60	19	25	30/3/25F/WP	HDB-43/45/47
RTU-D116	COMPUTER LAB	460/3/60	16	20	30/3/20F/WP	HDB-49/51/53
RTU-D117	SOCIAL STUDIES	460/3/60	24	30	30/3/30F/WP	HDB-55/57/59
RTU-D119	PRODUCTION LAB	460/3/60	14	20	30/3/20F/WP	HDB-61/63/65
RTU-D123	ART	460/3/60	29	35	60/3/35F/WP	HDB-67/69/71
RTU-D202	SPANISH	460/3/60	19	25	30/3/25F/WP	HDB-73/75/77
RTU-D206	RESTROOM	460/3/60	12	15	30/3/15F/WP	HDB-79/81/83
RTU-D209	SPANISH	460/3/60	19	25	30/3/25F/WP	HDB-44/46/48
RTU-D210	SPANISH	460/3/60	14	20	30/3/20F/WP	HDB-50/52/54
RTU-D211	SPANISH	460/3/60	16	20	30/3/20F/WP	HDB-56/58/60
RTU-D212	SPANISH	460/3/60	12	15	30/3/15F/WP	HDB-62/64/66
RTU-D213	SPANISH	460/3/60	16	20	30/3/20F/WP	HDB-68/70/72

FIGURE 9.9 Rooftop Electrical Schedule

LIGHTING FIXTURE SCHEDULE

TYPE	DESCRIPTION
A	RECESSED 2x4 FLUORESCENT LIGHT FIXTURE, ACRYLIC PRISMATIC DIFFUSER, STEEL DOOR FRAME, DIE-FORMED STEEL HOUSING, BAKED WHITE ENAMEL FINISH, ELECTRONIC BALLAST AND (4) F32/T8 LAMPS LIGHTOLIER #XT2GV1432-277-04
AX	SAME AS TYPE "A" EXCEPT WITH EMERGENCY BATTERY PACK CONTROLLING 2-LAMPS LIGHTOLIER #XT2GV1432-277-04-EM
B	SAME AS "A" EXCEPT WITH (2) F32/T8 LAMPS LIGHTOLIER #XT2G V1232-277-SO
BX	SAME AS TYPE "B" EXCEPT WITH EMERGENCY BATTERY PACK CONTROLLING 2 LAMPS LIGHTOLIER #XT2GV1232-277-SO-EM
C	RECESSED 2x4 - 32 CELL PARABOLIC LIGHT FIXTURE, SEMI-SPECULAR ALUMINUM LOUVER, DIE-FORMED STEEL HOUSING, BAKED WHITE ENAMEL FINISH, 2 ELECTRONIC BALLAST AND (4) F32-T8 LAMPS LIGHTOLIER #DPS2G32LS432-277-04
CX	SAME AS TYPE "C" EXCEPT WITH EMERGENCY BATTERY PACK CONTROLLING 2 LAMPS LIGHTOLIER #DPS2G32LS432-277-04-EM
D	SAME AS TYPE 'C' EXCEPT WITH (2) F32/T8 LIGHTOLIER #DPS2G32LS232-277-SO
DX	SAME AS TYPE 'D' EXCEPT WITH EMERGENCY BATTERY PACK LIGHTOLIER #DPS2G32LS232-277-SO-EM
E	4' WALL MOUNTED FLUORESCENT LIGHT FIXTURE, ACRYLIC LENS, STEEL HOUSING, ELECTRONIC BALLAST AND (2) F32-T8 LAMPS LIGHTOLIER #LSW240-277-SO
F	SURFACE MOUNTED 1'x4' FLUORESCENT STRIP, DIE FORMED STEEL HOUSING, BAKED WHITE ENAMEL FINISH, ELECTRONIC BALLAST, CHAIN HANGER SET, WIREGUARD AND (2) F32-T8 LAMPS. LIGHTOLIER # SW240RS-277-SS-AWG3-AH5
FX	SAME AS TYPE 'F' EXCEPT WITH EMERGENCY BATTERY PACK CONTROLLING 2 LAMPS. LIGHTOLIER # SW248SL-277-SS-AWG3-AH5-EM
G	3' FLUORESCENT STAGGERED STRIP, STEEL HOUSING, BAKED WHITE ENAMEL FINISH, ELECTRONIC BALLAST AND (2) F28-T8 LAMPS LIGHTOLIER #STG230-RS-HPF-277-SO
H	4' FLUORESCENT UNDER-COUNTER LIGHT FIXTURE, ACRYLIC LENS, STEEL BODY, WHITE FINISH, HPF BALLAST, 120V AND (1) F32-T8-LAMP LIGHTOLIER #TCU40W-120
J	400 WATT PENDANT MOUNTED METAL HALIDE, PRISMATIC BOROSILICATE GLASS REFLECTOR, MAGNETIC REGULATOR BALLAST, PORCELIAN LAMPHOLDER (LAMP INCLUDED) METAL GRILL, SAFETY CHAIN AND (1) MH400/HBU/PS PULSE START METAL HALIDE LAMP HOLOPHANE #PRSL-400MH-27-PD-E31-WG-212-B
K	RECESSED 70W. METAL HALIDE DOWNLIGHT, CLEAR ALZAK REFLECTOR, DIE-CAST ALUMINUM LAMPHOLDER AND HOUSING, THERMALLY PROTECTED, DAMP LOCATION LISTED AND (1) 70W ED-17-MH LAMP LIGHTOLIER #1113HR-1102HF/MHM75
K1	SAME AS TYPE "K" EXCEPT WITH (1) 100W-ED17 LIGHTOLIER #1113HR-1102HF/MHM100
L	DECORATIVE WALL SCONCE, METAL HOUSING, PERFORATED ACCENT PLATE, TEXTURED PAINT FINISH, ELECTRONIC BALLAST AND (1)CF26Q LAMPS ADVENT#AS4010-4011-CF26Q-12-277
M	WALL MOUNTED 175 WATT METAL HALIDE BOROSILICATE GLASS LENS, ALUMINUM HOUSING AND REFLECTOR, WET LOCATION LISTED AND (1) 175-ED17-MH LAMP EXCELINE #313-175-MAL-8
N	STAGE LIGHTING SYSTEM BORDER LIGHTS, REFER TO SPECIFICATIONS FOR ADDITIONASL INFORMATION. PROVIDE ALL LAMPS AS REQUIRED.
P	STAGE LIGHTING SYSTEM SOURCE 4 PAR SPOTLIGHTS, REFER TO SPECIFICATIONS FOR ADDITIONAL INFORMATION. PROVIDE ALL LAMPS AS REQUIRED.
R	RECESSED METAL HALIDE WALL WASHER, CLEAR ALZAK REFLECTOR, STEEL HOUSING, HIGH POWER FACTOR BALLAST, THERMALLY PROTECTED, DAMP LOCATION LISTED AND (1) 70W-ED17 MH LAMP LIGHTOLIER #WW7-CLW-WW7AC-70HC-F2
SA	(2) 400 WATT METAL HALIDE AREA LIGHTS ON 25'-0" SQUARE STEEL POLE, DIE CAST ALUMINUM HOUSING, TEMPERED GLASS LENS WITH ALUMINUM FRAME, TYPE III DISTRIBUTION, FINISH AND (2)MH400/C/U LAMPS EXCELINE #SMR4003MA-DB-480 KW INDUSTRIES #SSP25-4.0-7-DB-DM2180-BC
SB	SAME AS TYPE "SA" EXCEPT WITH TYPE II DISTRIBUTION AND (2) 250 WATT (WITH MH250/C/V LAMPS) METAL HALIDE FLOODLIGHTS MOUNTED ON BULLHORN EXCELINE #SMR4002MA-DB-480 WIDELIGHT#EFIM250HF-277-M2-DB KW INDUSTRIES #SSP25-4.0-7-DB-DM2180-BC-FS2180-2-DB
SC	SAME AS TYPE "SA" EXCEPT WITH (1) 400 WATT AREA LIGHT AND FLOODLIGHTS EXCELINE #SMR4003MA-DB-480 WIDELIGHT#EFIM250HF-277-M2-DB KW INDUSTRIES #SSP25-4.0-7-DB-DM10-BC-FS2180-2-DB
SD	SAME AS TYPE "SB" EXCEPT WITH (1)400 WATT AREA LIGHT AND FLOODLIGHTS EXCELINE #SMR4002MA-DB-480 WIDELIGHT#EFIM250HF-277-M2-DB KW INDUSTRIES #SSP25-4.0-7-DB-DM10-BC-FS2180-2-DB
SE	SAME AS TYPE "SA" EXCEPT WITH (1) 400 WATT AREA LIGHT AND HOUSE SIDE SHIELD EXCELINE #SMR4003MA-DB-480-SMRRGS KW INDUSTRIES #SSP25-4.0-7-DB-DM10-BC
SF	SAME AS TYPE "SB" EXCEPT WITHOUT FLOODLIGHTS EXCELINE #SMR4002MA-DB-480 KW INDUSTRIES #SSP25-4.0-7-DB-DM2180-BC
SG	SAME AS TYPE 'SB' EXCEPT WITH TYPE III DISTRIBUTION EXCELINE #SMR4003MA-DB-480 WIDELIGHT #EFIM250HF-277-M2-DB KW INDUSTRIES #SSP25-4.0-7-DB-DM2180-BC-FS2180-2-DB
SH	SAME AS TYPE 'SE' EXCEPT WITHOUT HOUSESIDE SHIELD EXCELINE #SMR4003MA-DB-480 KW INDUSTRIES #SSP25-4.0-7-DB-DM10-BC
X	SINGLE FACE LED EXIT LIGHT, BRUSHED ALUMINUM FACE-PLATE W/RED LETTERS, ULTRA THIN DIE CAST ALUMINUM CONSTRUCTION, HINGED FACE PLATE, AND UNIVERSAL MOUNTING 277V. LIGHT ALARMS # XLED-8-ARB
X1	SAME AS TYPE 'X' EXCEPT DOUBLE FACE LIGHT ALARMS #XLED-2B-ARB
X2	SAME AS TYPE 'X' EXCEPT WITH WIREGUARD LIGHT ALARMS # XLED-8-ARB-WG

FIGURE 9.10 Lighting Schedule

LIGHTING ELECTRICAL DISTRIBUTION SYSTEMS

When developing the electrical lighting loads for a commercial building, the engineer will first determine the lighting levels needed for each point in the building and for each activity being conducted. Having determined this, he can select luminaires, or fixtures, to provide the levels of lighting needed. In some cases where there are sophisticated lighting schemes or the lighting is to play an important part in the performance of the building, a lighting consultant may be used by the consulting electrical engineer.

Much of what we learned about identifying circuits for electrical power distribution also applies to lighting circuits. Notice in Figure 9.6 that the lighting circuit information is on its own plan view to avoid confusion with the power distribution circuits. At coordinates' locations E.4 and F along 18, we can see the lighting layout for the gym as before. We can now see that the gym lighting consists of J and FX light fixtures. These fixtures are switched at the entrance to the room. All of the lights for the gym are indicated by the arrow to be in panel H1. Notice how many light fixtures can be operated on the single 20-amp circuit and that the circuit nomenclature is the same as that used on the power distribution drawings. This similarity in identifying circuits provides speed for the electrical in identifying circuits.

The two types of lighting fixtures used in the gym are the J and FX—the difference, among other things, being that the FX fixture has a battery for operation if the power fails. Many other type of lighting fixtures are used in a commercial building, as indicated on the plan view. Again, a schedule is used, similar to Figure 9.10, to identify and provide nomenclature for each light fixture. In the schedule, an A fixture is identified as a 2′ × 4′ fluorescent light fixture, Lightolier Model #XT2GV1432-277-04-EM, operating at 277 V. Obviously, this is an excellent example of the use of a schedule to avoid putting too much information on the plan view.

SUMMARY

The electrical systems are the systems that give life to the components and activities in the building. The need for lighting is obvious, and most activities in a modern commercial building require various levels of power for support equipment. But even with the electrical system complete, we are not finished with looking at all the requirements for the project systems to work. We need to see how the plumbing category connects systems next.

10

PLUMBING CATEGORY OF THE DRAWINGS

INTRODUCTION

Plumbing systems comprise supply piping that distributes fluids or substances to a point of use, or collection systems that are used to capture any fluids or substances that are not consumed at the point of use. Many of these systems are needed for mechanical systems, so they will function properly. The piping and plumbing systems will carry an intermediate medium that the mechanical system will use for its process. Some of the typical plumbing symbols for these systems are found in Figure 10.12. We will talk about these plumbing and piping systems as supply piping systems and plumbing collection systems.

SUPPLY PIPING SYSTEMS

Piping systems in commercial buildings are primarily used to distribute fluids to some point of consumption, or point of use. There are many different types of fluids that can be used in a facility, depending on the business being conducted at that location. A manufacturing facility will use many more types of fluids than a school might use. The materials used as the piping system for each fluid is determined by the characteristics of the fluid and what type of pipe material is compatible with it. The pipe materials will be noted in the specifications, and the pipe routings will be shown in the drawings. As you look through the plans, you will notice that there are very few dimensions provided for locating pipes. Unless there is a specific reason that the engineer wants a pipe

Print and Specifications Reading for Construction, Updated Edition. Ron Russell.
© 2024 John Wiley & Sons, Inc. Published 2024 by John Wiley & Sons, Inc.
Companion website: www.wiley.com/go/printspecreadingupdatededition

RELATED DIVISIONS

Division 15
(Old Format)

Division 21 and 22
(New Format)

in a specific location, typically none will be provided at all. Usually, the drawings will only indicate a desired general routing for piping. It will be left to the general contractor's on-site superintendent to provide coordination of the crafts to avoid conflicts with building installations and components while ensuring that the systems function properly. Because little else is indicated in the drawings, it is important that the specifications be reviewed completely to understand the installation requirements for each system. There you will find the directions for assembling and installing the pipes. Even though there are many possible piping systems that can be used in a commercial building, there are some common systems that we should look at, starting with the piping used to deliver heated and chilled water to the fan coil unit 25 that we discussed in Chapter 8 regarding HVAC systems.

Heated/Chilled Water

FCU-25 was used to circulate the air into the building space. As we discussed in HVAC, this unit only moves the air through the supply ducts into the space and back to the unit through the return. This unit has coils inside that allow an intermediate material to be used to heat and cool the air. That intermediate material is water that has been heated or chilled. On the plan view in Figure 10.1, you can see FCU-25's location within the building and the pipes drawn to carry the water to the unit. These main water lines service many fan coil units, and taps are made into the pipes to allow each unit to connect to the system. The controls of the unit allow either the heated or chilled water to circulate through the coils, depending on what the thermostat is telling the controls that the temperature in the room is. The heated or chilled water is carried to the fan coils via the pipes—typically, black iron for this type of installation—from a boiler system and chiller plant, similar to those discussed in Chapter 8, Figure 8.1. Notice in Figure 10.1 that the pipes sizes are indicated in approximate locations where the engineer will allow a reduction in size. This usually is allowed as the pipes get farther from the source and the demand is for less volume of the fluid being supplied. The reductions in pipe sizes also allow for cost savings since smaller-diameter size pipe cost less. Each system, heated and chilled water, has a supply and return line. Even though the heated water has released its heat into the air, it still has energy in it and is warmer than fresh domestic make-up water. Therefore, we want to return this water to the boiler system so that it will take less energy to reheat it, as opposed to fresh domestic water. The same thought process is used with the chilled water.

In Figure 10.2, we can see a typical detail that would be shown in the drawings for the installation of the boiler system. In this particular example, two boilers are used for capacity and backup. The same is true for the chiller detail shown in Figure 10.3. In both of these examples, domestic water is supplied to the boilers or chillers and circulated through the piping system supply side using in-line pumps. The same pressure created by the pumps forces the water back to the units for reconditioning and recirculation. Both of these systems utilize domestic water as their process water, which is supplied by its own piping system.

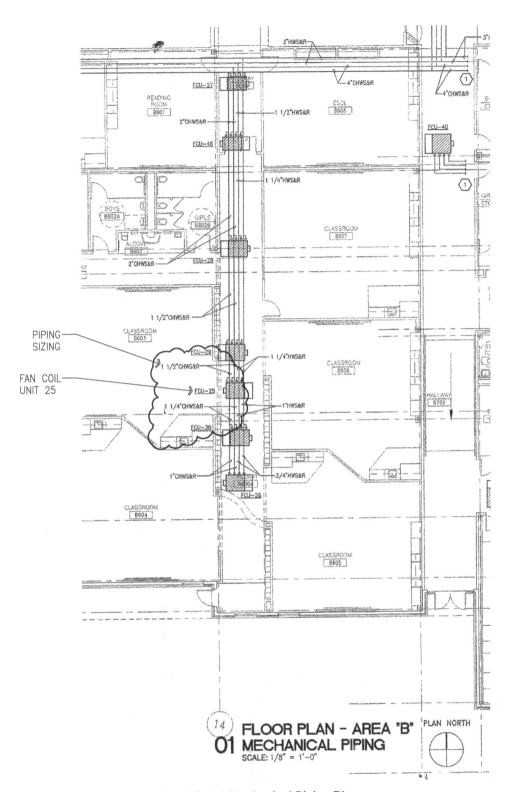

FIGURE 10.1 Mechanical Piping Plan

Process Water/Gases

As already discussed, a commercial building can have many types of process fluids and gases. Some of the most common include domestic water, natural gas, compressed air, and nitrogen. Medical or manufacturing facilities

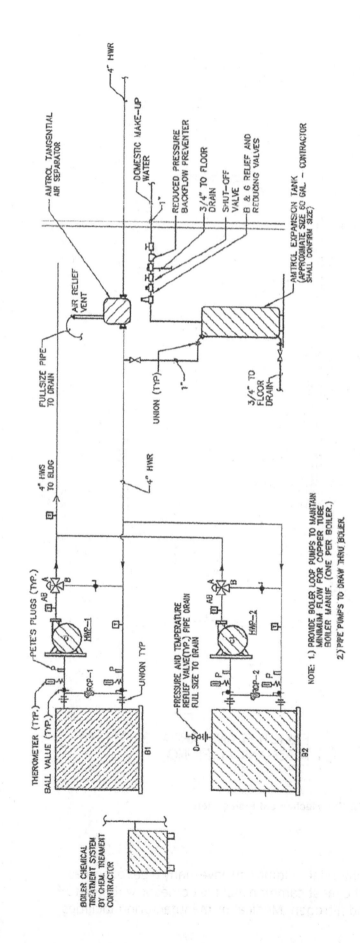

12 BOILER DETAIL

SCALE: NOT TO SCALE

FIGURE 10.2 Typical Boiler Detail

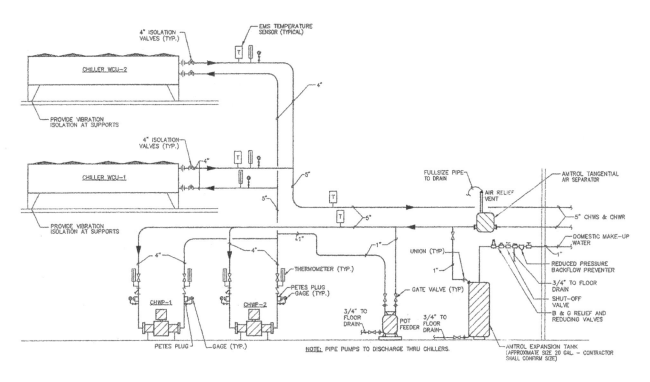

16 AIR COOLED CHILLER DETAIL
SCALE: NOT TO SCALE

AIR COOLED RECIPROCATING CHILLER

MARK#	NOMINAL TONNAGE BY MANUF.	WATER FLOW GPM	LWT °F	AMB TEMP °F	COOLER P.D. FT.	VOLTS PHASE	COMPR. RLA	COND. FAN FLA	MCA	MOCP	MANUFACTURER	MODEL	APPROX OPERATING WEIGHT (lbs.)	REMARK
WCU-1	110	203	42	102	6.0	460/3	1-117 1-92	3-3.8 3-3.8	262	300	YORK	YCAR0110SC	10151	SEE BELOW
WCU-2	110	203	42	102	6.0	460/3	1-117 1-92	3-3.8 3-3.8	262	300	YORK	YCAR0110SC	10151	SEE BELOW

NOTES:

1.) CHILLER MANUFACTURER SHALL PROVIDE FLOW SWITCH FOR EACH MACHINE. MECHANICAL CONTRACTOR SHALL INSTALL FLOW SWITCH, AND CONTROLS CONTRACTOR SHALL WIRE.

2.) PROVIDE 2 STAGES OF UNLOADING FOR EACH PART LOAD COMPRESSOR.

3.) PROVIDE FAILURE MODE ANALYSIS AT CHILLER CONTROL PANEL WITH MICROPROCESSED BASED ELECTRONIC CONTROLS.

4.) PROVIDE R22 REFRIGERANT MACHINES WITH SEMI-HERMETIC RECIPROCATING COMPRESSORS.

5.) PROVIDE STANDARD NON CORROSIVE COIL PROTECTION WITH HAIL GUARDS.

6.) PROVIDE STANDARD 0° OPERATION COND. FAN OPERATION WITH COOLER HEATER OPTION.

7.) PROVIDE CONTROL CIRCUIT POWER 30A @ 115V/1ø TO MACHINE WITH COOLER HEATER.

8.) PROVIDE MINIMUM CLEARANCES BETWEEN MACHINES, OTHER MACHINES, AND BUILDING AS REQUIRED BY CHILLER MANUFACTURER.

9.) PROVIDE CONCRETE HOUSEKEEPING PADS APPROXIMATELY 4" THICK. SEE STRUCTURAL/ARCH. DRAWINGS FOR DETAILS.

10.) PROVIDE RETURN RESET OPTION MODULE.

11.) CONTRACTOR TO PROVIDE ISOLATION VALVES AT EACH CHILLER. SEE DETAIL & CONTROLS FOR OPERATION.

12.) EACH MACHINE SHALL BE PROVIDED WITH 2 INDEPENDENT REFRIGERANT CIRCUITS.

13.) PROVIDE MINIMUM FLOW RATE INTERCONNECT WITH RESPECTIVE PUMP.

14.) CARRIER IS AN APPROVED ALTERNATIVE MANUFACTURER. MCQUAY IS NOT ALLOWED TO BID.

FIGURE 10.3 Chiller Details

can have many more types, depending on the processes preformed at the facility. Each of these process fluids requires a type of piping that is compatible with that fluid and may have requirements for installation that include special welding or mounting procedures that will be described in the specifications. If we look at the plan view in Figures 10.4 and 10.5, we can see how the piping routings are typically shown for most commercial buildings.

Unless there is requirement for a line to be in an exact location, the engineer will indicate the general routing and location with straight lines and allow the contractor to select the exact location and route for the pipes. Notice how the lines are labeled with the type of process fluid being transmitted and that the pipe size for each run is shown with the label. This is quite typical on drawings for most commercial buildings. On Figure 10.5, notice the natural gas lines. On Figure 10.4, notice the callout for Note 6. This note describes the routing of the domestic water line and its mounting requirements for the icemaker. Note 9 does this for the dishwasher shown in the same wall. The gas lines on Figure 10.5 show Note 4 and reference their intended consumption or distribution point, such as RTU2. The indicated routing of the lines, notes on the plan views, and requirements in the specifications all must be considered to completely understand the installation requirements for these process piping lines.

Figure 10.4 shows us another plan view of process piping and another type of callout that is commonly used. Look at the area with the arrow called Riser Diagram—Reference Figure 10.7. There you can see the callouts for the dishwasher, disposal, and sink. The definitions for these fixtures can be found in the specifications and typically on a schedule drawing similar the one in Figure 10.6, which would be found in the plans. The mounting heights and details can be found in the architectural category of the drawings. But how is the piping in the wall routed? In the callout area in Figure 10.4, there is a symbol that is a circle divided in half with a P in the top half and a 4 in the bottom half. This symbol is used to refer the drawing reader to a riser diagram that will detail the piping configuration for that group of fixtures. The reader would turn in the drawings and find a detail that looks like Figure 10.7. A riser diagram is a drawing that is used to show the schematic routing of the piping necessary at all elevations to connect all the fixtures indicated by the callout. The P/11 callout includes the hand sink, dishwasher, and disposal. In Figure 10.7, you can see those fixtures indicated and the specific pipe routings necessary to properly connect them, as well as the pipe sizes needed. The riser diagram provides clarification of the installation of these fixtures that cannot be shown on the plan view. We will come back to this type of drawing again in our discussion of plumbing, but now let's discuss one more typical piping system in a commercial building.

Fire Protection

This piping system will typically not be shown in the working drawings other than to indicate the locations of the sprinkler heads on the reflected ceiling plan for coordination with the lights and HVAC grilles and diffusers,

FIGURE 10.4 Piping Plan View

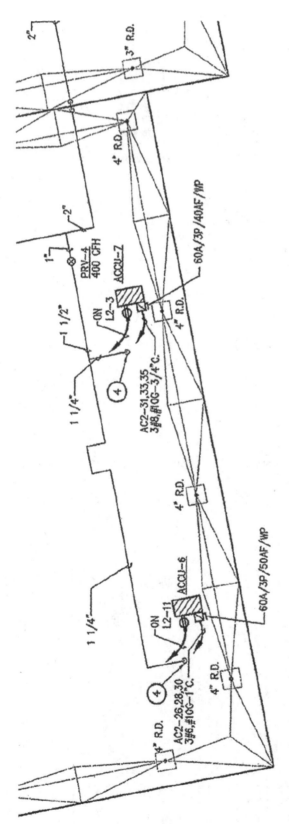

01 ROOF PLAN – M.E.P.
SCALE: 1/16" = 1'-0"

PLAN NORTH

FIGURE 10.5 Roof Piping Plan

PLUMBING FIXTURE SCHEDULE

FIXTURE	MANUFACTURER	REMARKS
*WC-1	AMERICAN STANDARD	FIXT. TYPE: WALL MTD. FLUSH VALVE MODEL NO: AFWALL 1.6 GPF #2257.103 SEAT: OLSONITE #95-OPEN FRONT, LESS COVER FLUSH VALVE: SLOAN #RESS-XD-C-1.6 BATTERY POWERED CARRIER: J.R. SMITH #410 SERIES COMPACT SUPPORTS.
*WC-2	AMERICAN STANDARD	FIXT. TYPE: WALL MTD. FLUSH VALVE MODEL NO: AFWALL 1.6 GPF #2257.103 SEAT: OLSONITE #95-OPEN FRONT, LESS COVER FLUSH VALVE: SLOAN #RESS-XD-C-1.6 BATTERY POWERED CARRIER: J.R. SMITH #410 SERIES COMPACT SUPPORTS.
*WC-3	AMERICAN STANDARD	FIXT. TYPE: WALL MTD. FLUSH VALVE MODEL NO: AFWALL 1.6 GPF #2257.103 SEAT: OLSONITE #95-OPEN FRONT, LESS COVER FLUSH VALVE: SLOAN #RESS-XD-C-1.6 BATTERY POWERED CARRIER: J.R. SMITH #410 SERIES COMPACT SUPPORTS.
*UR-1	AMERICAN STANDARD	FIXT. TYPE: WALL MTD. FLUSH VALVE MODEL NO: WASHBROOK 1.0 #6501.010 FLUSH VALVE: SLOAN #RESS-U-1.0 BATTERY POWERED CARRIER: J.R. SMITH #0635
UR-2	AMERICAN STANDARD	FIXT. TYPE: WALL MTD. FLUSH VALVE MODEL NO: WASHBROOK 1.0 #6501.010 FLUSH VALVE: SLOAN #RESS-U-1.0 BATTERY POWERED CARRIER: J.R. SMITH #0635
L-1	INTERSAN	FIXT. TYPE: WALL MOUNTED MODULAR, LAVATORY SYSTEM MODEL NO.: INTERSAN SANIFOUNT-MC-FOURSTATION WASHFOUNTAIN SHALL BE EQUIPPED WITH INTERSAN'S PASSIVE DETECTION SYSTEM. THE WASHFOUNTAIN SHALL INCLUDE A 10.8V "C" BATTERY PACK, STOP VALVES WITH A FLEXIBLE SINGLE SUPPLY WATER CONNECTION, AUTOMATIC FLOW CONTROL NOZZLE, LIMITING WATER CONSUMPTION TO LESS THAN .5 GPM. WATER FLOW SHALL SHUT OF 2.5 SECONDS AFTER REMOVAL OF HANDS FROM THE DETECTION FIELD WITH A MAXIMUM FLOW PERIOD OF 30 SECONDS.
*L-2	AMERICAN STANDARD	FIXT. TYPE: WALL-HUNG WHITE, FOR CONCEALED ARM SUPPORTS MODEL NO: LUCERNE #0355.012, ON 4" CENTERS FAUCET: CHICAGO #802-A-369 W/LEVER HANDLES TRAP: McGUIRE #8088 DRAIN: McGUIRE #155-WC, OFFSET GRID STRAINER SUPPLIES: McGUIRE #167-LK LOOSE KEY STOP LAV-GUARD: TRUEBRO #102 W/ACCESSORY #105
S-1	ELKAY	FIXT. TYPE: SS, SINGLE BOWL, 18 GAUGE-TYPE 302- SELF RIM MODEL NO: #LRAD-1919-55-1, CENTER HOLE FAUCET: CHICAGO #350 WITH LEVER HANDLE (GOOSENECK) TRAP: McGUIRE #8089 DRAIN: McGUIRE #1151-WC, OFFSET WITH BRASS STRAINER SUPPLIES: McGUIRE #2165LK, LOOSE KEY STOP EYEWASH: BRADLEY MODEL #S19-465EFW, DECK MOUNTED PULL OUT HAND HELD EMERGENCY EYEWASH. MOUNT SINK WITH FAUCET ON SIDE TO MEET TAS REACH REQUIREMENTS.
S-2	ELKAY	FIXT. TYPE: SS, DOUBLE BOWL, 18 GAUGE-TYPE 302- SELF RIM MODEL NO: DLR-3322-10 FAUCET: MOEN #8720 W/SINGLE HANDLE, SPRAY HEAD & HOSE TRAP: McGUIRE #8912 W/#111 WASTE PIPE SUPPLIES: McGUIRE #2165LK, LOOSE KEY STOP DISPOSER: BADGER2, 1/2 HP, 115 VOLT
S-3	ELKAY	FIXT. TYPE: SS, SINGLE BOWL, 18 GAUGE-TYPE 302- SELF RIM MODEL NO: LRAD-1522-55-4 FAUCET: MOEN #8720 W/SINGLE HANDLE, SPRAY HEAD & HOSE TRAP: McGUIRE #8089 SUPPLIES: McGUIRE #2165LK, LOOSE KEY STOP LAV-GUARD: TRUEBRO #102 W/ACCESSORY #105
S-4	ELKAY	FIXT. TYPE: SS, SINGLE BOWL, 18 GAUGE-TYPE 302- SELF RIM MODEL NO: LRAD-1919-55-1, CENTER HOLE. FAUCET: CHICAGO #350 WITH LEVER HANDLE (GOOSENECK) TRAP: J.R. SMITH #5730 PLASTER TRAP DRAIN: McGUIRE #1151WC OFFSET WITH BRASS STRAINER SUPPLIES: McGUIRE #2165LK, LOOSE KEY STOP MOUNT SINK WITH FAUCET ON SIDE TO MEET TAS REACH REQUIREMENTS.
S-5	ELKAY	FIXT. TYPE: SS, SINGLE BOWL, 18 GAUGE-TYPE 302- SELF RIM MODEL NO: DLR-1919-10, CENTER HOLE FAUCET: CHICAGO #350 WITH LEVER HANDLE (GOOSENECK) TRAP: McGUIRE #8912 W/ #111 WASTE PIPE SUPPLIES: McGUIRE #2165LK, LOOSE KEY STOP
FD-1	J.R. SMITH	FIXT. TYPE: DUCO CAST BODY WITH FLASHING COLLAR AND ADJUSTABLE STRAINER HEAD MODEL NO: #2005A-P050
FD-2	J.R. SMITH	SAME AS FD-1 EXCEPT WITH A 6" FUNNEL
RPBFP	WATTS	FIXTURE TYPE: REDUCED PRESSURE BACKFLOW PREVENTER VALVE ASSEMBLY MODEL NO.: 909 PROVIDE ISOLATION VALVES AND UNIONS. RP2 SHALL BE LOCATED IN A BELOW GRADEHOT BOX REFER TO CIVIL DRAWINGS FOR LOCATION.
HB	WOODFORD	FIXT. TYPE: WALL MOUNTED HOSE BIB W/ VACUUM BREAKER MODEL NO: #24P WITH TEE KEY
TP	J.R. SMITH	FIXT. TYPE: AUTOMATIC TRAP PRIMER MODEL NO: #2699
*EWC-1	ELKAY	FIXT. TYPE: WALL MTD. ELECTRIC WATER COOLER MODEL NO: LZS-8 MOUNTING HEIGHT PER ADA/TAS CAPACITY: 8.0 GALLONS PER HOUR ELECTRICAL: 120V, 1 PHASE SUPPLIES: McGUIRE #167LK, LOOSE KEY STOP
*EWC-2	ELKAY	FIXT. TYPE: WALL MTD. ELECTRIC WATER COOLER MODEL NO: LZS-8 CAPACITY: 8.0 GALLONS PER HOUR ELECTRICAL: 120V, 1 PHASE SUPPLIES: McGUIRE #167LK, LOOSE KEY STOP
WH-1	A.O. SMITH	FIXT. TYPE: GAS FIRED STORAGE WATER HEATER MODEL NO: BTR-500, 500,000 BTUH, 85 GALLON STORAGE 473 GPM RECOVERY AT 100°F. RISE. PROVIDE WATTS DETA-12 EXPANSION TANK.
WH-2	A.O. SMITH	FIXT. TYPE: WALL MTD. ELECTRIC WATER COOLER MODEL NO: #DSED-30, 30 GAL. STORAGE, 41 GPH RECOVERY AT 90°F RISE, 9 KW ELECTRICAL: 480V, 3 PHASE
WH-3	A.O. SMITH	FIXT. TYPE: WALL MTD. ELECTRIC WATER COOLER MODEL NO: #DSED-10, 10 GAL. STORAGE, 13 GPH RECOVERY AT 90°F RISE, 3 KW ELECTRICAL: 277V, 1 PHASE
RCP-1	BELL & GOSSETT	FIXTURE TYPE: IN-LINE RECIRCULATING PUMP MODEL NO: B&G SERIES #100 32 GPM @ 8'-0" TOTAL HEAD ELECTRICAL: 115V., 1 PHASE
TV	WATTS	FIXTURE TYPE: HOT WATER TEMPERING VALVE SET FOR 105°F MODEL NO.: L70A

FIGURE 10.6 Plumbing Fixture Schedule

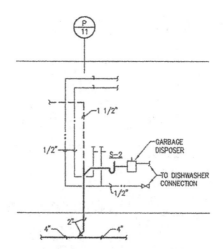

FIGURE 10.7 Plumbing Riser Diagram

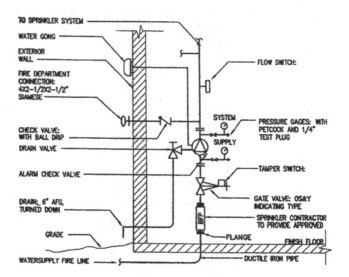

FIGURE 10.8 Fire Riser

and in the mechanical and civil details. These drawings show only the fire protection line from the domestic water main to the building and how that line is distributed in the building at the riser system using support equipment such as the pumps shown in Figures 10.8 and 10.9. The fire protection distribution system must be engineered by someone licensed in fire protection systems, who will also create the shop drawings. This is usually done by the subcontractor who will be doing the work at the construction site. Once the subcontractor has completed the shop drawings, they will be submitted to the general contractor (who will provide copies to the architect) and usually the local municipality's fire marshal (who will review and approve the plans for construction). Once approved, these shop drawings become part of the contract documents and the system

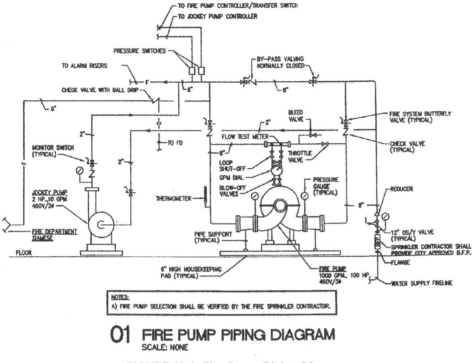

01 FIRE PUMP PIPING DIAGRAM
SCALE: NONE

FIGURE 10.9 Fire Pump Piping Diagram

must be installed per those plans. The definitions of the materials that can be used and how the piping system is installed will be defined in the specifications.

PLUMBING COLLECTION SYSTEMS

As we discussed before, commercial buildings can have many types of piping systems for process fluids. These piping systems deliver the fluids to a point of use or consumption. If the fluids delivered are not totally consumed at this point of use, then measures must be taken to collect the excess fluids and either treat them before disposal or allow them to be disposed of directly to a municipal sewer system. This piping system must also have a vent system that will allow the gases from the wastes in the piping to escape into the atmosphere to avoid potential explosions in the system. Many of these collection piping systems will have compatibility requirements the same as the process piping systems, depending on the nature of the fluid being collected. Again, most of these requirements will depend on the actual processes within the facility and what type of business is being conducted. Many manufacturing and medical facilities will have many waste stream that cannot be combined due to reactive properties in the materials. Many of these waste streams must be treated to neutralize the materials before they can be disposed of to a sanitary sewer. However, most commercial buildings that simply conduct business with minimal processes will have some basic plumbing requirements that we will discuss here.

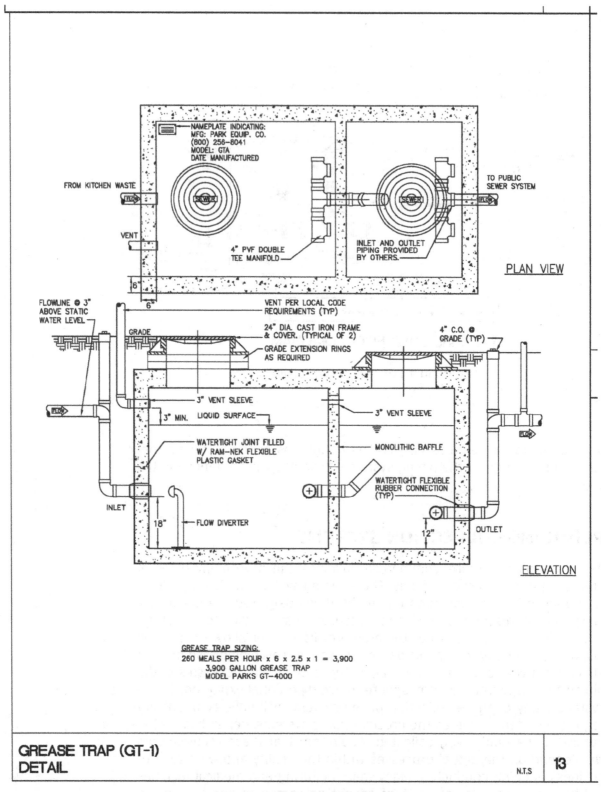

PLAN VIEW

NAMEPLATE INDICATING:
MFG: PARK EQUIP. CO.
(800) 256-8041
MODEL: GTA
DATE MANUFACTURED

FROM KITCHEN WASTE

TO PUBLIC
SEWER SYSTEM

VENT

4" PVF DOUBLE
TEE MANIFOLD

INLET AND OUTLET
PIPING PROVIDED
BY OTHERS.

6"

6"

FLOWLINE @ 3"
ABOVE STATIC
WATER LEVEL

VENT PER LOCAL CODE
REQUIREMENTS (TYP)

GRADE

24" DIA. CAST IRON FRAME
& COVER. (TYPICAL OF 2)

GRADE EXTENSION RINGS
AS REQUIRED

4" C.O. @
GRADE (TYP)

3" VENT SLEEVE

3" MIN. LIQUID SURFACE

3" VENT SLEEVE

WATERTIGHT JOINT FILLED
W/ RAM-NEK FLEXIBLE
PLASTIC GASKET

MONOLITHIC BAFFLE

WATERTIGHT FLEXIBLE
RUBBER CONNECTION
(TYP)

INLET

18"   FLOW DIVERTER

12"   OUTLET

ELEVATION

GREASE TRAP SIZING:
260 MEALS PER HOUR x 6 x 2.5 x 1 = 3,900
3,900 GALLON GREASE TRAP
MODEL PARKS GT-4000

GREASE TRAP (GT-1)
DETAIL

N.T.S

13

FIGURE 10.10 Typical Floor Drain

Sanitary Sewer

In Figure 10.4, we looked at a plan view that showed us an area where a hand sink, dishwasher, and disposal are to be installed. In Figure 10.7, we saw the riser diagram that indicated the piping system needed to deliver the domestic water to these fixtures. Now that we have a source of water that will not be totally consumed, we must have a collection system for that source. In Figure 10.7 we can see the riser diagram for that collection system. Notice that this riser diagram has the same callout as the diagram in both the supply piping and collection system in P/11. This makes it easy in the plans to identify the supply piping and plumbing systems that correspond to the same fixture installations. This diagram shows the hand sink with a 2-inch drain connecting it to a 4-inch main drain line. A typical floor drain is shown in Figure 10.10. The effluent from these combines and flows into a larger 6-inch line that will ultimately flow into the city sanitary sewer system. It also shows the $1\frac{1}{2}$-inch vent line that would extend out the roof to allow gases to escape. For those fluids that cannot be discharged directly to the sanitary sewer, some measures must be taken to treat the effluent before discharge, such as water contaminated with oils or grease.

Grease Traps

When oils and greases are mixed with water, they cannot be directly discharged to the municipal sewer systems. An interceptor called a grease trap is used in the plumbing line to allow the oils and grease to be captured and separated from the waters so the water can be discharged to the sewer. Figure 10.11 shows a typical grease trap.

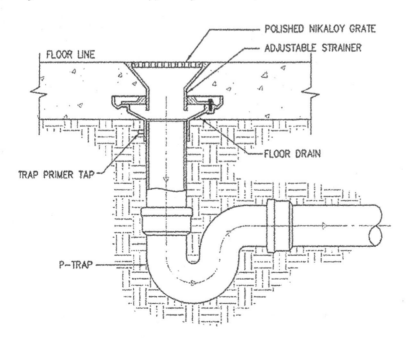

FIGURE 10.11 Grease Interceptor Detail

PLUMBING SYMBOL LEGEND

SYMBOL	DESCRIPTION
————————	SANITARY WASTE PIPING
— — — — —	SANITARY VENT PIPING
—— S ——	STORM DRAIN PIPING
—— OF ——	STORM DRAIN OVER FLOW PIPING
—— • ——	COLD WATER PIPING
—— • • ——	HOT WATER PIPING
—— • • • ——	HOT WATER RECIRCULATION PIPING
—— D ——	CONDENSATE DRAIN LINE PIPING (D) (CD)
—— G ——	NATURAL GAS PIPING
⌐⊕	REGULATOR
▼	GAS COCK
◁▷	GATE VALVE
●	BALL VALVE
⊢/↗	CHECK VALVE
⊣|⊢	UNION
◼	BALANCING VALVE
▶	FLOW DIRECTION OF PIPING
◁⊗▷	SOLENOID VALVE
⊕	HOSE BIBB OR NON FREEZE WALL HYDRANT (HB) (NFWH)
⊏—	CAP PIPING
▢	FLOOR DRAIN (FD)
◎	HUB DRAIN (HD)
▩	FLOOR SINK (FS)
⊙	FLOOR CLEANOUT (FCO)
⊙⊙	TWO WAY OR DOUBLE CLEAN OUT (TWCO) (DCO)
⊩—	WALL CLEANOUT (WCO)
VTR	VENT THROUGH ROOF
UF	UNDER FLOOR
OH	OVERHEAD ABOVE CEILING
◕	POINT OF CONNECTION TO EXISTING
◗	SEE CIVIL DRAWINGS FOR CONTINUATION

ALL SYMBOLS ON THIS LIST ARE NOT NECESSARILY USED ON THIS JOB

FIGURE 10.12 Plumbing Symbols

Drains from selected waste streams suspected of containing oils or greases would be routed to the trap, where they would be forced through filter media, typically sand. The oils would be trapped and the clean water allowed to discharge to the sanitary sewer system. Other waste streams with contaminates that cannot be discharged directly to the sewer system would have similar or even more sophisticated systems. Usually, the local municipality will require registration and permitting of these types of treatment systems.

SUMMARY

As already noted, there can be many different types of plumbing and piping systems that can be installed in a modern commercial building. Most of the common ones are discussed in this chapter. These systems will allow the building to be occupied by the people who will be conducting the business at that facility. As varied as these systems are, the symbols that will be used are even more diverse. In Figure 10.12, we can see the typical symbols used in the plumbing category of the drawings for the plumbing and piping components. The drawing symbols used for a particular building may be different in the next one because each building can have different requirements.

Much of the building project—including mechanical, electrical, and plumbing—can lie outside the confines of the structure on the site. We will address this next as we explore the civil work or sitework.

11

CIVIL/SITEWORK CATEGORY OF THE DRAWINGS

OVERVIEW

Up to this point, we have studied elements of the drawings that have pertained to the building interior or the building itself. A large portion of the project will also include the site on which the project is built and any improvements to be made. There are three categories in the commercial drawing set that we must look at to see this work: civil, architectural, and mechanical, electrical, and plumbing (MEP).

The civil drawings will contain most of the information directly related to the site, or property improvements, including the changes in property grading, any paving, and basic utility services to the property. The architectural drawings will contain information regarding site signage for traffic and parking, and the definition of any other project elements being constructed on the site. The MEP drawings will contain information defining the connection of the building utilities to the site utilities.

All three of these categories must be looked at in order to completely understand the entire scope of work for the site itself. Let's examine these more closely to see what is contained in these drawings.

RELATED DIVISIONS

Division 2 (Old Format)

Division 31, 32, 33, 34, and 35 (New Format)

Print and Specifications Reading for Construction, Updated Edition. Ron Russell.
© 2024 John Wiley & Sons, Inc. Published 2024 by John Wiley & Sons, Inc.
Companion website: www.wiley.com/go/printspecreadingupdatededition

CIVIL DRAWINGS

As already stated, this category of drawings will contain the most information directly related to the property itself, as well as its definition. This category will also define the utilities that are required to connect the building to the various utility suppliers. We start at the beginning with the definition of the property.

Plat Drawings

Some of the first drawings found in the civil category will be the plat drawings. These drawings define the property boundaries with metes and bounds using coordinates, direction, and distance from a known benchmark to define the property lines. This drawing will also show proposed improvements and utility easements for the new project. There will also be a written description of the property boundaries with the drawings. These drawings will be created by a licensed surveyor and filed with the appropriate jurisdiction, city, county, or state. This definition of the property boundaries will allow the civil engineer to begin his work on the site plan.

Site Plan

This drawing will provide the definition of the site improvement areas, including buildings, paving, drives, playground areas, walkways, park or garden areas, and some basic references to the utilities on the site. This drawing is used to coordinate all the different elements being added to the site. The architect's site plan will be the basis for this drawing, as we will discuss later. This drawing will also contain the site benchmarks used to locate everything added to the site and will establish the datum for the project. The datum usually references the building's first-floor elevation. This elevation is used as the datum to measure everything up or down vertically. Using the actual elevation would be cumbersome since it is seldom an even number, so the datum is established as 100'-0", which makes calculations simpler. The site plan contains all of the dimensions needed to locate the project elements horizontally, as well. These dimensions control the layout of the site horizontally from the benchmarks and property lines. If the project site is very complicated or congested, these dimensions may be placed on a separate drawing called the dimension or horizontal control drawing. Once we have the locational definition of the site, the civil engineer can begin to define the site contours.

Grading Plan

The civil engineer will, with consideration of existing structures and proposed additions to the site, develop the grading of the site. On a drawing that shows the site boundaries, the engineer will draw the contour of the site, using lines to identify each change of elevation in feet across the site. Figure 11.1 illustrates a sample from a grading plan. The existing contour lines indicate that the site is currently 584 feet above sea level and are the lighter ones. These are existing contour lines showing the site as it is now. The proposed contour lines would put the finished floor at 588 feet

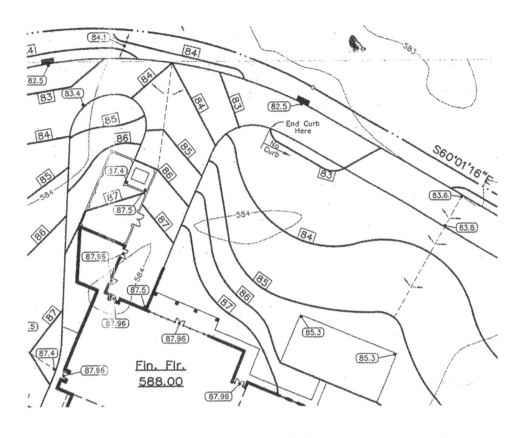

Proposed Spot Elevation

Existing Spot Elevation

Proposed Contour

Existing Contour

Direction Of Flow

Drainage Divide

BENCH MARK

City of Murphy Bench Mark, square cut in center of headwall, west side of F.M. 2551, 900' north of Tom Clevenger Road.

Elevation: 576.00

FIGURE 11.1 Grading Plan View

above sea level and are indicated by the darker lines. This suggests that the site should be raised by 4 feet. The contractor will use these contour lines to determine how to move the soils on the site to meet the requirements for the project. These contours will also be used to establish grades that will allow for controlling drainage on the site.

Drainage Plan

The grading plan indicates the general flow of surface waters across the site. In many cases, these waters can be allowed to flow naturally into a drainage ditch or other waterway such as a creek or lake. Each site is contoured so that waters can be controlled and not cause flooding on the project site or adjacent properties. Development of this drawing is often done in conjunction with the U.S. Army Corps of Engineers to ensure that the area waterways are not compromised. To assist in controlling the flow of water, a drainage plan is developed to indicate which direction waters will flow across the site and where water will need to be collected and removed from the site using drains to storm sewers or waterways.

Storm Drains

Figure 11.2 shows the same plan view as Figure 11.1. In Figure 11.1 the contour lines showed the site basically draining in a direction toward the top of the plan view. To facilitate the removal of water from the site, Figure 11.2

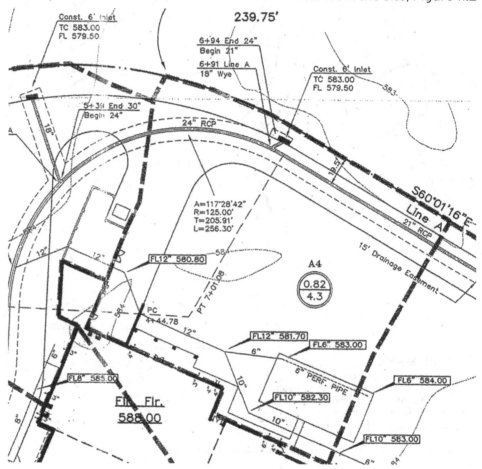

DRAINAGE CRITERIA

Q = C I A
C = 0.7
C = 1.0 Roof
I_{100} = 7.52
tc = 15 min.

XX — Drainage Area Number

X.XX — Acres
X.X — Q_{100} (cfs)

Drainage Divide Line

DRAINAGE AREA CALCULATIONS

DRAINAGE AREA No.	DRAINAGE AREA (Acres)	C	t_o (min)	I_{100} (in/hr)	Q_{100} (cfs)
A1	0.26	0.7	15	7.52	1.4
A2	0.21	0.7	15	7.52	1.1
A3	0.80	0.7	15	7.52	4.2
A4	0.82	0.7	15	7.52	4.3
A5	1.37	0.7	15	7.52	7.2
B1	0.69	0.7	15	7.52	3.6
B2	0.99	0.7	15	7.52	5.2
B3	0.67	0.7	15	7.52	3.5
C1	0.10	0.7	15	7.52	0.5
C2	0.10	0.7	15	7.52	0.5
D	0.61	0.7	15	7.52	3.2
E	0.42	0.7	15	7.52	2.2
F	0.23	0.7	15	7.52	1.2
R1	0.72	1.0	15	7.52	5.4
R2	0.26	1.0	15	7.52	1.9
R3	0.29	1.0	15	7.52	2.2
R4	0.08	1.0	15	7.52	0.6
R5	0.16	1.0	15	7.52	1.2
R6	0.23	1.0	15	7.52	1.7

FIGURE 11.2 Drainage Plan

shows a drainage line with curb inlets at the top of the plan going around the site. This drainage line is designed of reinforced concrete pipe, (RCP), and is sized to remove a calculated volume of water from the site.

This line (see Figure 11.3) will prevent the accumulation of water on the property and avoid wash areas that would result from allowing the water to wash naturally over the embankment.

This example shows the sewer line in section view, as though a section has been cut through the property at the drainage line. This drawing is not called a section view, though; it is called a profile. This profile drawing shows the line in section so the viewer can see the changes in elevation from one end of the line to the other. The profile indicates the slope of the line, as well as start-and-stop points for each elevation change.

Erosion Control/Stormwater Pollution Prevention

Construction sites today must have a stormwater pollution prevention plan developed specifically for their site and approved by the local governing EPA region. The intent is to prevent construction debris, material, and hazardous materials from entering into national waterways from the construction site. Figure 11.4 shows the same plan view of the storm drain, with inlet control shown. The inlet control would be similar to that shown in Figure 11.5, and specifically as prescribed in the stormwater pollution prevention plan. The intent, of course, is to block solid debris from the entrance of the storm drain while allowing the water to filter through and out to the waterway. In addition to this, the stormwater pollution prevention plan will specify other means of preventing construction debris from entering waterways. The most prevalent of these today is the silt fence for erosion control. As shown in Figure 11.6, the typical silt fence is designed to stop soil erosion from leaving the site and entering area storm drains adjacent to the site.

Paving Drawings

With the grades established and the erosion control in place, work can now start on the site, and one of the first construction activities to begin is the site paving. The plan views will show the areas to be paved, as defined on the site plan, and will include the construction, expansion, and control joints. These joints are indicated on the plan view and coordinated because they are visible and must also be located to minimize cracking and breaking of the paving. Figure 11.7 shows the difference between these joints. The expansion joint is designed and placed to allow the pavement to expand when hot without buckling or breaking. The control joint is a sawed joint placed on a specific spacing to allow the paving to contract without cracking, or if it cracks it will crack in a controlled manner—that is, along the sawed joint. The construction joint location is determined by the contractor and is designed to allow the contractor to develop a stopping point for construction. Usually, the contractor will select or plan a location that corresponds with a control joint to maintain the paving pattern design.

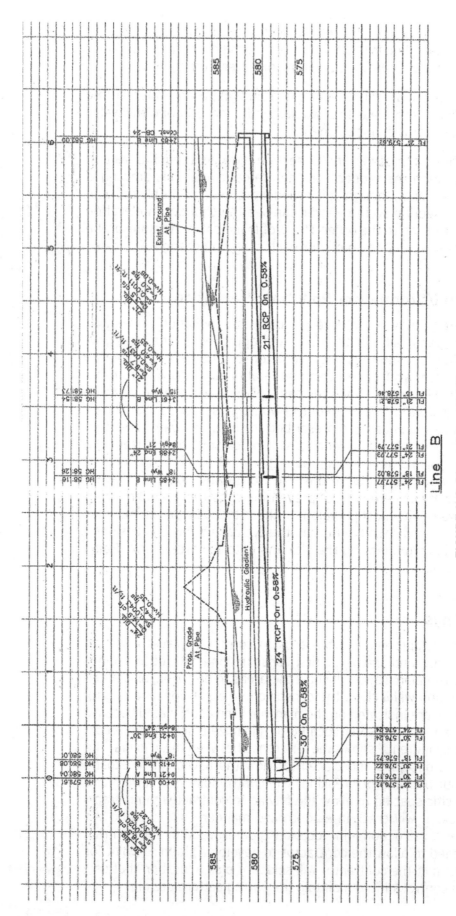

FIGURE 11.3 Storm Sewer Profile

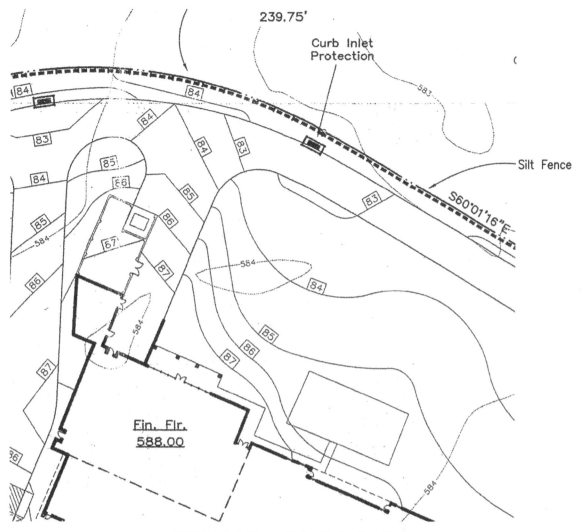

FIGURE 11.4 Storm Water Protection Plan

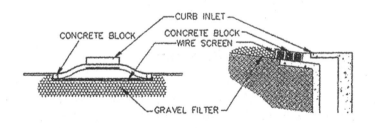

CURB INLET PROTECTION

FIGURE 11.5 Curb Inlet Protection

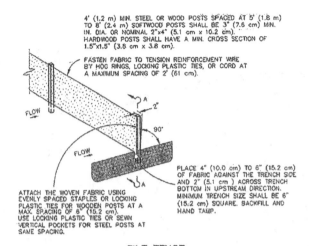

4' (1.2 m) MIN. STEEL OR WOOD POSTS SPACED AT 5' (1.8 m)
TO 8' (2.4 m) SOFTWOOD POSTS SHALL BE 3" (7.6 cm) MIN.
IN. DIA. OR NOMINAL 2"x4" (5.1 cm x 10.2 cm).
HARDWOOD POSTS SHALL HAVE A MIN. CROSS SECTION OF
1.5"x1.5" (3.8 cm x 3.8 cm).

FASTEN FABRIC TO TENSION REINFORCEMENT WIRE
BY HOG RINGS, LOCKING PLASTIC TIES, OR CORD AT
A MAXIMUM SPACING OF 2' (61 cm).

FLOW

2'

90'

FLOW

PLACE 4" (10.0 cm) TO 6" (15.2 cm)
OF FABRIC AGAINST THE TRENCH SIDE
AND 2" (5.1 cm) ACROSS TRENCH
BOTTOM IN UPSTREAM DIRECTION.
MINIMUM TRENCH SIZE SHALL BE 6"
(15.2 cm) SQUARE. BACKFILL AND
HAND TAMP.

ATTACH THE WOVEN FABRIC USING
EVENLY SPACED STAPLES OR LOCKING
PLASTIC TIES FOR WOODEN POSTS AT A
MAX. SPACING OF 6" (15.2 cm).
USE LOCKING PLASTIC TIES OR SEWN
VERTICAL POCKETS FOR STEEL POSTS AT
SAME SPACING.

SILT FENCE

FIGURE 11.6 Silt Fence Detail

Water and Sanitary Sewer Plans

With most of the project grade changes established, the site utilities can now be defined. The domestic water and sanitary sewer lines will be defined from where they connect to the closest mains from the city, up to the point where they will enter the building. The domestic water line will show the piping from the connection at the city main to where it is split for fire protection service into the building and the irrigation system, which will be detailed in the landscaping drawings. The main lines for these utilities are shown here because the sitework contractor is usually the same contractor that installs the underground piping for these systems. The final connection for the piping going into the building will be shown in the MEP drawings.

Site Detail Drawings

The site detail drawings will contain all of the details needed for the paving and sidewalks, drainage connections to the stormwater drains, site signs, and any other site additions, such as retaining walls, gates, or other structures.

Landscape and Irrigation Drawings

Most modern commercial construction projects in a municipality will require a certain amount of landscaping to be included in the project, and that municipality will require that the landscaping be irrigated to maintain appearances. Usually the architect will have a landscape architect develop the plans for the site, which will include the irrigation. The irrigation plans must be done by someone licensed to design, and sometimes install, these types of systems using the landscape architects plans. The landscaping requirements will be provided by the governing municipality and the landscape architect will include those requirements in the plans. Figure 11.8 shows what these requirements might be and how they were addressed. Also shown in this example is a schedule for the plants and trees used. Figure 11.9 provides some samples of planting details that would be included in the drawings by the landscape engineer.

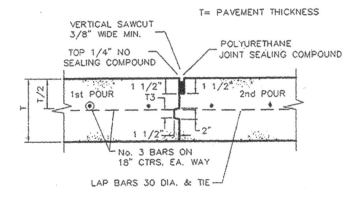

CONSTRUCTION JOINT
N.T.S.

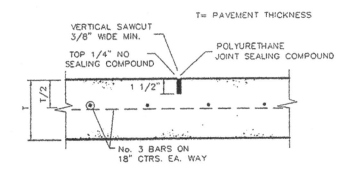

SAWED DUMMY (CONTROL) JOINT
N.T.S.

MAXIMUM SPACING IS 15' CTRS. (TYP.)

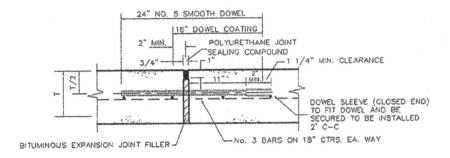

EXPANSION JOINT
N.T.S.

FIGURE 11.7 Paving Joints

LANDSCAPE TABULATIONS

LANDSCAPE EDGE
1-3" TREE PER 50 LF, 10-5 GAL. SHRUBS PER 500 SF WHERE PKG OR DRIVES ABUTS OR BERM 18 TO 40 INCHES, IF PKG. MORE THAN 50' AWAY THEN NO SCREEN

LANDSCAPE EDGE AREA (1,036 LF FRONTAGE X 10 =)	10,360 SF
LANDSCAPE EDGE TREES REQUIRED (10,360 SF \ 500 SF = 20.7)	21 TREES
LANDSCAPE EDGE TREES PROVIDED	21 TREES
LANDSCAPE SCREEN REQUIRED (SHRUBS OR BERM)	SHRUBS OR BERM
LANDSCAPE SCREEN PROVIDED	BERM

INTERIOR PARKING
8 SF PER PARKING SPACE, 1-3" TREE PER 15 SPACES

TOTAL PARKING SPACES	103 SPACES
INTERIOR PARKING LANDSCAPE AREA	
PARKING LANDSCAPE AREA REQUIRED (103 X 8 SF = 824 SF)	824 SF
PARKING LANDSCAPE AREA PROVIDED	2,751 SF
INTERIOR PARKING TREES (3" CAL. MIN.)	
PARKING TREES REQUIRED (103 SPACES / 15 = 6.9)	7 TREES
PARKING TREES PROVIDED	7 TREES

EXISTING TREES
REPLACE INCH PER INCH OF PROTECTED TREES

PROTECTED TREES REMOVED (24,24,18 = NOT PROT./BLDG. PAD)	66 CAL. INCHES
REPLACEMENT TREES PROVIDED (31 X 3" =)	93 CAL. INCHES

ALL LANDSCAPE AREAS TO RECEIVE AUTOMATIC UNDERGROUND IRRIGATION

SCHOOL PLANT LIST

TREES

QUANTITY	SYMBOL	CALLOUT	COMMON NAME	SCIENTIFIC NAME	SIZE & CONDITION
10	⊕	LIVE OAK	Live Oak	Quercus virginiana	3" caliper, 10'-12' Ht./ 4'-5' spread, B&B straight trunk
7		SWEET GUM	Sweet Gum	Liquidambar styraciflua	3" caliper, 10'-12' Ht./ 4'-5' spread, B&B straight trunk
7		CADDO MAPLE	Caddo Maple	Acer barbatum	3" caliper, 10'-12' Ht./ 4'-5' spread, B&B straight trunk
7		RED OAK	Standard Red Oak	Quercus shumardii	3" caliper, 10'-12' Ht./ 4'-5' spread, B&B straight trunk
14		TREE YAUPON	Yaupon Holly	Ilex vomitoria	1" caliper per trunk, 5 trunk minimum, 9'-11' Ht./9' spread, B&B or container, heavily berried specimen, limbed to 4', full all sides

SHRUBS

QUANTITY	SYMBOL	CALLOUT	COMMON NAME	SCIENTIFIC NAME	SIZE AND CONDITION
3	Ⓒ	PHOTINIA	Frasers Photinia	Photinia x Fraseri	15 gallon 6' Ht./36" spread, bushy, full to ground
242	⊙	D.B. HOLLY	Dwarf Burford Holly	Ilex cornuta 'Burfordi nana'	5 gallon, 18"-20" Ht./15" spread, full, bushy to ground
11	·	D. BARBERRY	Crimson Pygmy Barberry	Berberis thunbergi 'Crimson Pygmy'	3 gallon, 12" Ht./14" spread full, bushy to ground
88	·	D.Y. HOLLY	Dwarf Yaupon Holly	Ilex vomitoria 'Nana'	5 gallon, 16" Ht./14" spread, bushy, full to ground

GROUNDCOVER / VINES

QUANTITY	SYMBOL	CALLOUT	COMMON NAME	SCIENTIFIC NAME	SIZE AND CONDITION
AS SHOWN		LIRIOPE	Liriope	Liriope muscari 'Spicata'	4" pots, 8" o.c. full
AS SHOWN		GRASS	Common Bermuda Grass	Cynodon dactylon	Hydromulch refer to specifications
AS SHOWN		SOD	Common Bermuda Grass	Cynodon dactylon	Sod refer to specifications. A single row of sod shall be placed along the edge of any pedestrian paving that is adjacent to a grass area shown to be hydromulch.

MISCELLANEOUS

QUANTITY	SYMBOL	CALLOUT	COMMON NAME	SCIENTIFIC NAME	SIZE AND CONDITION
AS SHOWN	——	STL. EDGE	L.F. Ryerson steel edge 1/8" x 3" with 12" stakes, green in color		

FIGURE 11.8 Landscape Schedules

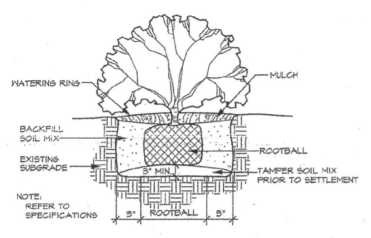

WATERING RING

MULCH

BACKFILL SOIL MIX

EXISTING SUBGRADE

ROOTBALL

TAMPER SOIL MIX PRIOR TO SETTLEMENT

NOTE: REFER TO SPECIFICATIONS

3" MIN.

3" ROOTBALL 3"

SHRUB PLANTING DETAIL (TYPICAL)

SCALE: N.T.S.

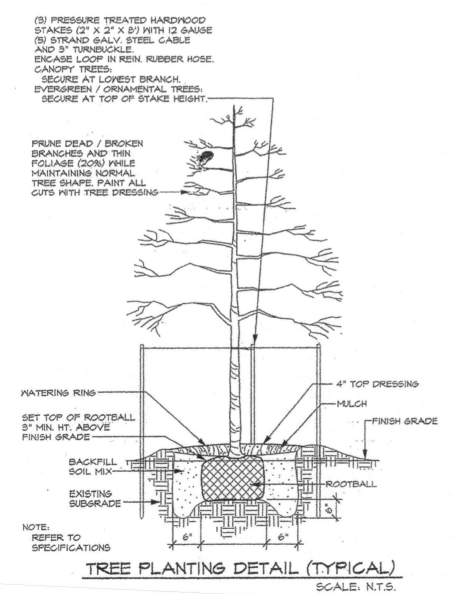

(3) PRESSURE TREATED HARDWOOD STAKES (2" X 2" X 8') WITH 12 GAUGE (5) STRAND GALV. STEEL CABLE AND 3" TURNBUCKLE.
ENCASE LOOP IN REIN. RUBBER HOSE.
CANOPY TREES:
  SECURE AT LOWEST BRANCH.
EVERGREEN / ORNAMENTAL TREES:
  SECURE AT TOP OF STAKE HEIGHT.

PRUNE DEAD / BROKEN BRANCHES AND THIN FOLIAGE (20%) WHILE MAINTAINING NORMAL TREE SHAPE, PAINT ALL CUTS WITH TREE DRESSING

WATERING RING

SET TOP OF ROOTBALL 3" MIN. HT. ABOVE FINISH GRADE

BACKFILL SOIL MIX

EXISTING SUBGRADE

4" TOP DRESSING

MULCH

FINISH GRADE

ROOTBALL

6"

NOTE: REFER TO SPECIFICATIONS

6" 6"

TREE PLANTING DETAIL (TYPICAL)

SCALE: N.T.S.

FIGURE 11.9 Landscape Details

All of the work included by the civil engineer in the civil drawings requires close coordination with the project architect and consulting MEP engineers to ensure that the property meets the owner's requirements. It should also be noted that some local municipalities will provide their own construction details for some of the items found in the civil drawings. The civil engineer will typically include these standard city details in the drawings. This will help expedite the review of the project with the municipality because the city will see the details included as part of the project. Next we will look at the civil, or site, information provided by the architect.

ARCHITECTS SITE PLANS

In the architectural category of the drawings, the architect will include a site plan that will define the layout of the site, including parking areas, walls, fences, patios, walks, and flagpole locations. This layout will have been completed with the owner so the site can be maximized for the operations that will take place there and will provide dimensions for sizing and locating these additions on the site. The aesthetics of these project elements will be defined here as well, such as elevations of signs, gazebos, retaining walls, and so on.

The architect will also define any signage required for traffic and parking. Figure 11.10 shows a sample of how the architect determined how many accessible parking spaces were required and what the signage and parking space markings should look like. The architect will also define the striping for all drives, fire lanes, and general parking areas.

The civil and MEP engineers will use the site plan to engineer property grades and utilities that will allow the site to function. Now let's look at the MEP engineer's role in the site development.

MEP DRAWINGS

The consulting engineers who completed the drawings for the MEP categories will also complete the drawings for the site utilities. Since the main utility lines will be defined in the civil category, the consulting engineers will complete the definition of the utilities connections to the buildings, specifically for the sanitary sewer, natural gas, and domestic water, to include all piping and valve configurations for backflow prevention and separation.

The MEP drawings will also have the roof drain definitions for the roof plan and will show the connections to the storm sewer. This drawing will show the locations of the drains on the roof, how the lines are run, and where they continue to underground storm drains.

The MEP drawings will contain information for the site electrical distribution. It will show the power coming across the site to the site transformer and from there to the building switchgear, either overhead or underground depending on local requirements. The MEP electrical will also include the requirements

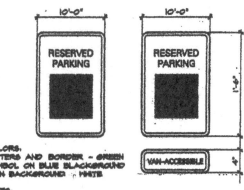

PARKING COUNT	
72	CAR SPACES PROVIDED
3	HANDICAPPED SPACES (VAN ACCESSIBLE)
75	TOTAL SPACES PROVIDED
72	PARKING SPACES REQUIRED 1:500 RATIO
2	HANDICAPPED SPACES REQUIRED
74	TOTAL SPACES REQUIRED
74	BUS PARKING SPACES

COLORS:
LETTERS AND BORDER – GREEN
SYMBOL ON BLUE BACKGROUND
SIGN BACKGROUND – WHITE

NOTES:

1. SPACING BETWEEN LETTERS, COLORS, AND PROCESSES SHALL CONFORM STANDARD HIGHWAY AND SIGN DESIGNS FOR TEXAS

2. INSTALL WHERE INDICATED ON PLANS

3. VAN ACCESSIBLE SIGNAGE ON VAN SPACES ONLY.

05 H.C. SIGN DETAIL
SCALE: N.T.S.

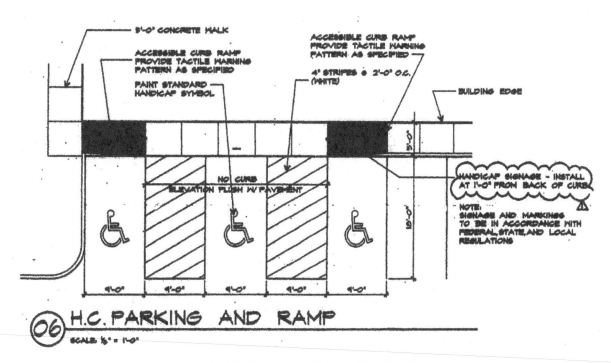

06 H.C. PARKING AND RAMP
SCALE: ¼" = 1'-0"

FIGURE 11.10 Accessible Parking Details

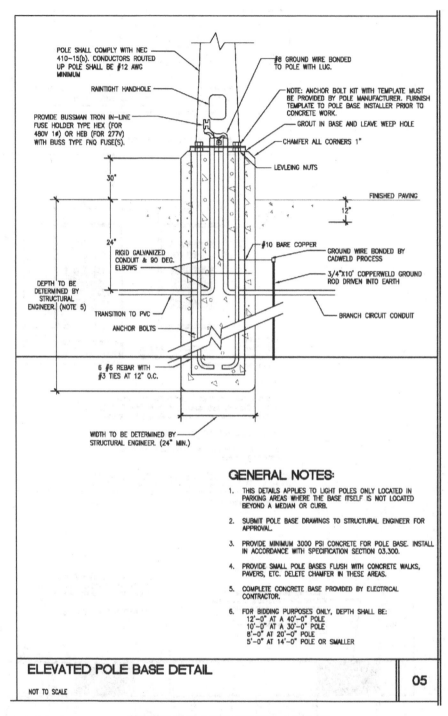

FIGURE 11.11 Light Pole Base Detail

for communications wiring, cabling, and fiber optics for phones, cable, CCTV, security, fire alarms, and other communications systems.

The site lighting requirements will also be on these drawings. The lighting circuits will be shown the same way circuits are shown in the electrical category of the drawings and will indicate the panel and breakers feeding the circuit, conduit sizes, and wire sizes. Any electrical circuits for site devices such as gate openers, lighted signs, and fountains would be indicated in the same method.

The MEP drawings will also provide any details needed for the connection of any of these utilities. A common detail is shown in Figure 11.11. This detail shows the level of coordination among the consulting engineers needed to complete the site definition. This light pole base detail has the electrical definition provided by the electrical consulting engineer, the concrete base provided by the structural engineer, and the installation of the base defined by the civil engineer, based on the site soil conditions.

As indicated with this detail, all of these consulting engineers and the architect must coordinate closely to accomplish the definition of the site for the owner.

SUMMARY

As we can see, to complete the site for the project, it requires the close coordination of the architect and the civil and MEP engineers. The changes to the property for the new construction must be completed before installation of the site utilities. The architect will work with the owner to define the site layout, and the consulting MEP engineers will complete bringing utilities to the building. The work in these drawings will complete the definition of the commercial construction project for the general contractor and subcontractors. The civil engineer will provide the definition of the changes to the property needed to prepare the site for the project based on the site plan developed by the architect with the owner. The MEP engineer will define the utilities needed for the project to function as required by the owner.

GLOSSARY

A

Accessible buildings Buildings that comply with the Americans with Disability Act.

Addendum A document issued by the architect prior to execution of a construction contract that modifies the construction drawings and specifications.

Agreement between owner and architect The contract between the owner and the architect for the design drawings and specifications, a standard American Institute of Architects publication.

Agreement between owner and contractor The contract between the owner and a general contractor for construction services, a standard American Institute of Architects publication.

Architect A licensed individual who develops the drawings and specifications for construction of a building.

Architectural drawings A category in the construction drawings that best describes the building; the term is occasionally used in reference to the entire drawing set for constructing a building.

B

Backer rod A compressible round plastic material that when pressed into the space in a construction or expansion joint provides a backing that prevents caulk or other sealant material from sagging and reduces the amount of sealant material used.

Print and Specifications Reading for Construction, Updated Edition. Ron Russell.
© 2024 John Wiley & Sons, Inc. Published 2024 by John Wiley & Sons, Inc.
Companion website: www.wiley.com/go/printspecreadingupdatededition

Benchmark A known point in elevation above sea level, longitude, or latitude use by a survey engineer to locate elements of a construction project.

Bidding To submit a quote or price for construction work, typically in competition with other contractors.

Boiler A mechanical device used for heating water for steam or hot water.

Bond beam A masonry course in a wall filled with grout and reinforcing to provide a horizontal beam in the masonry wall.

Bond pattern The pattern created when block or brick is laid in place in a masonry wall.

Bricks A masonry unit typically made of clay.

Built-up roofing systems A roofing system comprised of different layers of roofing felts with rock ballasting on top.

C

Callout symbols Symbols used by the architect or engineer to direct people around a drawing or from sheet to sheet to access the different views.

Categories of drawings Used to group drawings together that contain similar information such as civil, architectural, structural, mechanical, electrical, or plumbing.

Change order A modification to the construction contact for increased or decreased scope of work.

Chiller Mechanical devise for cooling air or water.

Civil drawings Drawings that contain most of the information directly related to the site or property.

Collection piping Plumbing piping that consists of the sanitary sewer, drains, and vents to the atmosphere.

Conception phase Phase of construction in which the owner decides that he needs a building and begins to develop or purchase land and hires an architect to begin designing the building.

Concrete blocks A masonry unit made from concrete and small rock, usually structural when applied to a masonry wall.

Construction management delivery method Delivery method in which the contractor manages several other contractors to accomplish the work but is not engaged in the work.

Construction manager One who manages construction work or projects for the owner.

Construction phase Phase during the project is built; final phase of construction.

Construction Specification Institute The institute that created a method for categorizing construction information into the divisions of the specification.

Consulting engineers Engineers that are hired by the architect to do the design work on the building systems, usually consisting of a civil, structural, mechanical, electrical, and plumbing engineer.

Contract documents   Documents that comprise a contract for construction consisting of as a minimum the agreement between the owner and architect, the agreement between the owner and contractor, the general conditions of the contract for construction, specifications, the working drawings, and any addendums.

Convenience outlet   Electrical outlet spaced as a convenience for the owner; usually there are several to a circuit.

Conveyances   Any number of methods for moving people or things a distance either horizontally or vertically such as elevators and escalators.

Cubic feet per minute (CFM)   Measure of volume of air moved by a fan or some air device.

Cutting plane   An imaginary plane used to indicate a location where a building section is taken from or drawn.

D

Dedicated circuit   Electrical outlet that has a single outlet in a specific location dedicated to a specific use or to a piece of equipment.

Design phase   The phase of construction where the owner and the architect work together and with others to complete a design of a building.

Design-build delivery method system   A single contract with one firm for the design and construction of the project.

Details   Part of a section drawing drawn at a larger scale to show greater detail.

Developer   A person who takes a raw piece of real estate and develops the infrastructure such as roads, fire protection, and other utilities.

Division sections   The succinct sections of a division, a division can consist of many sections depending on a buildings scope of work.

Divisions (specifications)   The 50 divisions create by the Construction Specification Institute to categorize construction information.

Door conventions   Used to determine the direction of swing, or the direction the door opens.

Drawings   The collection of all the categories of drawings for a building; also called sheets.

Duct   A conduit that conducts fluids or air; typically refers to air ducts in construction.

E

EIFS   External insulation and finish system composed of insulation, a lath system, and stucco applied as a finish.

Electric elevators   An elevator that can move quickly between many floors; also called a friction elevator due to the use of friction brakes.

Elevation   A type of drawing used to illustrate to exterior views of a building.

Escalators   A moving stairway that moves the rider from floor to floor.

Estimator Individual who is responsible for developing the costs and sometimes schedule for a construction project.

F

Fire protection piping A piping system designed to adequately provide water to suppress a fire.

Flashings Membranes or metals designed in conjunction with a roofing system to protect and cover fasteners and openings against water infiltration.

G

General conditions of contract for construction An American Institute of Architects publication with 14 Articles for the Administration of Construction Contracts.

General contractor The prime contractor on a construction project.

General, product, execution The three parts of any division sections.

Grade beams Beams that are placed at grade usually on piers.

Gravel guard A metal flashing at the end of a flat roof shaped so that the ballast does not wash off the roof.

Grease interceptor/trap An in-line device on a sewer line designed to allow the grease from kitchen operations to be separated from the rest of the sewage and collected at a later date.

Gypsum Board A drywall compound made into a panel or sheet and hung onto a metal frame.

H

Hydraulic elevators Elevators that are designed to carry heavy weights but do not cover a lot of floors.

Index of drawings A listing of all the drawings included in a drawing set typically on the first or second sheet.

I

Instructions to bidders The instructions to bidders, including when the bids are due, where to deliver the bids, and other details about bidding process.

K

Key words Words in a statement about construction that indicate whether the information being sought is in the drawings or specifications.

L

Line of sight The direction that a view originated from, indicated on the drawings in a callout.

M

MasterFormat® The 50 divisions developed by CSI.

Mechanical drawings   Drawings that contain heating, ventilating, air conditioning systems.

MEP   Mechanical, electrical, plumbing.

O

One-line diagram   An electrical drawing that shows the distribution of electricity throughout the building.

Open web joists (wood and metal)   Joists designed to span from beams and girders that have an web design that allows for duct, pipes, and conduits to be ran through the webbing or hung from the bottom cord.

Overhead door   A door or grille that either coils or is suspended from overhead inside the building.

Owner   The person who has the idea and money on construction projects.

P

Panelboard   An electrical device that contains breakers that shut off the power to individual circuits.

Phases of construction   Conception, promotional, design, construction.

Piers   A structural feature that supports the building by being drilled down to a solid substrate, usually rock, and comes up to support the building grade beams.

Pipe taps   A devise is clamped on a charged pipe and the pipe is tapped and threaded while charged, often done when pipes change direction.

Pitch pan   A metal pan constructed to allow pipe and conduit penetrations through a roofing system without leaking, usually filled with a tar substance or sealant.

Plan view   A drawing view of an object with a line of sight from above.

Plastic laminates   Layers of plastics pressed into a single surface used to provide a slick covering for counters and the front of cabinetry and millwork.

Plumbing drawings   Drawings that show piping supply systems and collection systems.

Plumbing riser diagrams   A drawing of a piping system that shows piping components in their correct order typically in the vertical plane.

Preliminary drawings   Drawings created for the owner to verify the design of the building.

Presentation drawings   Drawings or models of the construction project that are done to show what the projects real features will look like when completed.

Process piping   To supply piping distribution, usually to a device or a system.

Project manager   One who has the authority to manage and make decisions regarding the progression of a construction project.

Project manual   A manual for each construction project that contains the relative owner documents, contracts, bidding information, and specifications for that project.

Project team A team formed for the construction of a building project consisting of the owner, the architect, and the general contractor.

R

R value A materials resistance to heat flow, the higher the r-value a material has the better that material is at resisting heat flow.

Realtor A person who sells land and buildings that have been already developed with roads and infrastructure.

Record drawings/as-builts Drawings that are marked by the contractor that indicate any changes to the drawings regarding how a project was actually installed.

Reinforcing steel Metal steel bars used to reinforced concrete and give it tensile strength.

Resilient flooring Several flooring materials that give when walked upon generally placed on top of concrete.

Roll-up door/grille Similar to an overhead door, used to cordon off spaces inside a building; typically placed in a corridor.

Rooftop unit (RTU) Heating, ventilating, and air conditioning units that are complete heat and cooling units placed on the roof of a building.

S

Sanitary sewer Collection piping that consists of drains that carry away the sewage and waste from a building.

Schedule A series of tasks and dates used to complete work on construction projects.

Sealants Rubberized compounds used to seal joints and cracks in buildings.

Section (drawings) A drawing of a building indicated by a cutting plane and line-of-sight flag that shows material placement and details in side walls.

Sheets The individual drawings in a drawing package.

Shop drawings Drawings created by a subcontractor who is supplying a component or part of a building indicating what and how he is going to install, typically provided to the general contractor who provides them to the architect who reviews them.

Single-ply membrane roofing systems A roofing system comprised of a single ply of material, usually a plastic material.

Site grades Comprised of contour lines that indicate elevations above sea level used to establish the lay of the land.

Sketches Drawing used by the owner and architect to define a construction project.

Slab-on-grade Concrete slab, usually a flooring surface, that is poured directly on the ground.

Specifications Methods, materials, quality.

Steel shapes   The different shapes used to construct the structure of the building consisting of W shapes, S shapes, angles, channels, pipe, tubing, and plate.

Storefront system   A system of windows and doors made out of an extruded material, usually aluminum, constructed on site, often used for the entrance of a building.

Storm water pollution prevention plan   A plan devise by the contractor and approved by the federal government to prevent storm water from running off the construction site and carrying construction debris and soils into creeks and water ways.

Structural category/drawings   Drawing that show the structure created from the different structural shapes.

Subcontractor   A contractor who works for another contractor on a construction project.

Supply piping   Piping that supplies process fluids to equipment or systems.

Survey   Used to establish the boundaries of a property, benchmarks are used to establish the boundaries of a construction project.

Switchgear   An electrical device that usually contains the main switch for the electricity to a building.

T

Terrazzo   A flooring system created from small chips of colored rock that creates a hard shiny surface.

Traditional contract delivery method   The conventional bidding system for a construction project where there is a separate contract between the owner and architect and the owner and contractor.

U

Utility easement   A document that allows major utilities to run pipes, conduits, and overhead wires across a property.

V

Vinyl-clad tile   Resilient vinyl composition floor tile most often used to cover concrete.

W

Warranty   A guaranty provide by the contractor for workmanship and materials typically for a one-year period.

Working drawings   The drawings used to bid and build a building, provide size, shape, location.

INDEX

Print and Specifications Reading for Construction, Updated Edition. Ron Russell.
© 2024 John Wiley & Sons, Inc. Published 2024 by John Wiley & Sons, Inc.
Companion website: www.wiley.com/go/printspecreadingupdatededition